Experiments in Physical Optics

Experiments
in Physical Optics

M. Françon   **N. Krauzman**   **J. P. Mathieu**   **M. May**

University of Paris

GORDON AND BREACH SCIENCE PUBLISHERS

New York   London   Paris

Copyright © 1970 by:

Gordon and Breach, Science Publishers, Inc.
150 Fifth Avenue
New York, N.Y. 10011

Editorial office for the United Kingdom:

Gordon and Breach, Science Publishers Ltd.
12 Bloomsbury Way
London W.C. 1

Editorial office for France:

Gordon & Breach
7–9 rue Emile Dubois
Paris 14^e

Translated by S. Mallick, Institute of Optics, Paris

Preface

The experiments described in this book are not meant to be performed by the students, but are intended for class demonstration to illustrate the teaching of physical optics. Only a small number of these experiments are quantitative. Most of them are easily visible, directly or by projection, from all parts of a lecture theatre. The others can only be observed individually though most of these can be shown to the audience with the help of a television camera and screens.

In each case, we have briefly given the principle of the experiment, enumerated the required apparatus, given a diagram of the set-up and indicated how to make the necessary adjustments. Only a small number of optical pieces (lenses, prisms, mirrors, polarizers) is required to set up most of the experiments.

In certain cases it is difficult to avoid the use of commercial equipment. However, if the reader is familar with glass blowing and the assembling of electronic circuits, he will be able to set up some of the experiments described. We have not entered into the details of these subjects, and most of the components employed in these experiments in physical optics do not need a specialized workshop for their construction.

We shall be grateful to those of our readers who send us their suggestions and criticism.

M.F.   N.K.   J.P.M.   M.M.

Contents

1

Sources of Continuous and Line Spectra

1.1   LIGHT SOURCES

1.1.1   Carbon arc

The types of carbon arcs suitable for experiments in optics have a vertical carbon rod of 0.5 cm diameter and a horizontal one of 0.7 cm diameter (fig. 1.1). This latter must be connected to the positive terminal of the D.C. source; its crater C serves as the source in the H direction. It radiates

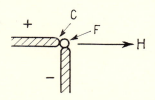

Figure 1.1

approximately as a black body at a temperature of 3,500 °C and consequently its emission spectrum is continuous. Its brightness is more than 10^8 nits. The arc is lighted by bringing the two carbon rods into contact and then separating them immediately by a distance of about one centimeter. The crater C must not be masked by the negative carbon rod in the horizontal direction.

The voltage applied to the arc is of the order of 45 volts; the current, normally 4 to 5 amps, may be increased for a short duration in order to increase the light intensity. The arc has a characteristic $V = f(I)$ with a negative slope. It is therefore essential, to obtain a stable arc that a D.C. voltage of 110 or 220 volts be applied and a rheostat with a maximum resistance of 20 Ω (110 V) or of 50 Ω (220 V) be connected in series with the arc.

If one wants to examine the radiation from the flame F of the arc, one should observe in a direction normal to the plane of the figure (1.1); or a carbon arc with vertical carbon rods (fig. 1.2) should be employed. In this case, the arc can be worked on alternating current.

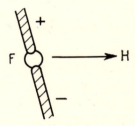

Figure 1.2

1.1.2   Arcs between metallic electrodes

1   Replacing the two carbon rods in figure 1.2 by iron rods of diameter 0.4 cm approximately and using a current of 2.5 to 3 A, we obtain a flame emitting a line spectrum; the spectrum consists of a large number of lines in the visible and the ultraviolet regions continuing up to 0.2 μ. These lines given in various tables (*) are used as secondary standards for the measurement of wave-lengths.

2   The arc between copper rods gives a spectrum which is also rich in lines.

3   The arc between tungsten electrodes, contained in a bulb filled with an inert gas, provides a source of very small dimensions (pointolite lamp), one of the electrodes being spherical in form; the brightness is of the order of 10^7 nits.

1.1.3   Incandescent lamps

1.1   Lamps with tungsten filaments in an atmosphere of inert gas are well known. Their brightness is of the order of 10^7 nits. The form given to the coiled filament is often unsuitable for experimental purposes. However,

---

* Atlas of Fabry and Buisson; Atlas of Vatican.

it is possible to obtain filaments in which the axis of the helix is a straight line.

1.2 Lamps with tungsten ribbon have a better geometrical shape. They work under a voltage of 6 to 10 V and with a current of 10 to 20 A. They are worked on mains with the help of a transformer.

1.3 Tube lamps with tungsten filament and with quartz envelope, working directly on the mains, have a low brightness but produce radiations of wavelengths up to about 1.3 μ. These are very suitable for use in near infra-red.

1.1.4 Gas discharge

1 Lamps containing sodium vapour produce a light which is sufficiently monochromatic for many purposes (line D_1, $\lambda_1 = 5896$ Å, line D_2, $\lambda_2 = 5890$ Å). Sodium being a solid at ordinary temperature and its vapour pressure being very small, the lamp contains argon which enables the discharge to pass between the electrodes and gives a small amount of pink light. The metal evaporates as the temperature rises due to the discharge and the yellow light of sodium starts appearing. After a few minutes this is the only light which remains visible.

2 The lamps containing other metals (K, Rb, Cs, Zn, Cd, Hg) function in an analogous manner.

All the preceeding lamps work on 110 V or 220 V through an auto-transformer supplied especially with the lamp. When the steady state is reached, the voltage across the terminals is of the order of 20 V and the current is of the order of 1 A.

3 Lamps containing mercury vapour which attain a pressure of ten atmospheres and have a quartz envelope, yield an intense source of radiations in the visible and the near ultra-violet. The principal lines are

λ(Å) 3022 3130 3655 4047 4358 5461 5780

They do not contain any rare gas. One of the electrodes E (fig. 1.3) is cold; the other E' is adjacent to an auxiliary electrode e put in series with a resistance R of 20,000 Ω contained in the lamp which works on 110 V or 220 V across a special inductance S. When the lamp is switched on, a discharge passes between E' and e and energizes the lamp; then, the current passes between E and E' on account of the high value of R. After a few minutes, the brightness reaches a value of the order of 10^7 nits.

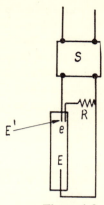

Figure 1.3

4   These lamps can be furnished with a glass envelope which suppresses ultra-violet.

5   On the other hand, they may have glass envelopes which are opaque to visible light and transparent to ultra-violet.

6   Mercury lamps in which the pressure rises to 200 atmospheres and which have to be cooled by circulating water, have a brightness approaching 10^9 nits; their spectrum is almost continuous.

High voltage (thousands or ten thousands of volts) discharge in an atmosphere of gas at low pressure (of the order of 10^{-3} atm.) yields two sources of secondary interest.

7   The hydrogen tube (Geissler tube) whose emission contains in particular the lines of the Balmer series.

8   The mercury tube with an envelope of quartz or of a special glass emits, under these conditions, very little radiation in the visible spectrum, but principally the resonance line ($\lambda = 2537$ Å).

1.1.5   Lasers

There exist in the market diverse models of lasers with He-Ne, which yield a parallel beam of coherent, monochromatic radiation $\lambda = 6328$ Å, carrying a flux of 5 to 50 mW according to the model. The weakest model suffices for the experiments described here.

In the usual experiments, it is sometimes necessary to enlarge the laser beam by means of a microscope objective. The beam coming out of the objective presents defects of uniformity due to the diffraction of the incident light by dust particles on the objective and by the laser itself. Sometimes they are very marked and may be troublesome for the observation of the phenomena. To eliminate these parasite diffractions which make large angles with the principal beam, it suffices to place in the image focus of the objective a diaphragm with a circular hole of about 25 µ, which lets pass only the diffraction disc of the laser. The laser beam being of smaller cross section than the pupil of the objective, one observes in the image focus the diffraction figure of the laser and not that of the objective.

1.2   OPTICAL FILTERS

Optical filters are used to transmit spectral bands of known width in a continuous spectrum or to isolate certain radiations in a line spectrum. They are based either on the selective absorption shown by certain chemicals (cf. chap. 20) or on the phenomena of constructive interference (cf. § 6.1.1)

1.2.1   Absorbing filters

There exist in the market* various glass or gelatin colour filters, which have the following uses:

— Absorption of the ultra-violet and transmission of the visible (colourless glasses)
— Transmission of the ultra-violet and absorption of the visible (Wood's glass, black glasses)
— Absorption of the infra-red and transmission of the visible (pale blue glasses)
— Transmission of the infra-red and absorption of the visible (black glasses)
— Neutral glasses attenuating white light without colouring it with a variable degree of general absorption (grey glasses).
— Numerous red glasses, receiving light from a carbon arc, transmit a sufficiently monochromatic light suitable for most of the experiments on interference and diffraction.

Besides the absorbing solid filters, one can use various solutions in cells with parallel faces. We give below the commercial names of glass filters and the chemical solutions suitable for isolating diverse lines of mercury.

---

* Schott, Wratten, Sovirel. The catalogues can be had on demand.

Transmitted radiations in Å	Schott	Wratten	Solutions in water
3650	UG 1	18A	
4358	CG 7	50	0.0075 g of rhodamine B in 100 cm³. Saturated sodium nitrite (in separate cells).
5461	BG 20 + CG 11	77	Potassium bichromate and saturated neodym nitrate (in separate cells)
5780	OG 2	22	Eosine or concentrated potassium bichromate

1.2.2   Interference filters

These are based on the principle of the Fabry–Perot interferometer (cf. §7.5.1), and are obtained by depositing on a dielectric plate reflecting films of metal or multilayers of some dielectric substance. Filters for isolating the visible lines of mercury, anti-caloric filters, semi-reflecting mirrors and cold mirrors are available commercially.

1.3   PROJECTION OF SPECTRA

1.3.1   Principle

The light radiation, the spectrum of which we want to obtain, is dispersed by a prism or a grating. A simple method for the projection of spectra consists in illuminating a slit F as intensely as possible with the light to be analysed, and to form an image of the slit by a converging system followed by the dispersing element.

On figure 1.4, the converging system is a lens L and the dispersing element is a prism P.

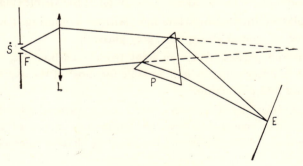

Figure 1.4

In figure 1.5, the converging system is a concave spherical mirror *M* and the dispersing element is a reflection grating *R*. In this arrangement, the light does not traverse the optical element and thus there is no risk of absorption.

Obviously, one can associate a prism with a mirror or a grating with a lens and employ in this later case, a transmission grating.

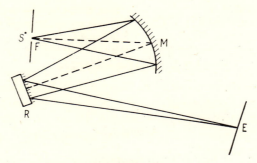

Figure 1.5

1.3.2 Setting up the experiments

It depends on the spectral region to be studied.

a) Apparatus

S: carbon arc for a continuous spectrum in the visible, infra-red and near ultraviolet. Mercury vapour lamp of A_3 type for a line spectrum in the visible and the ultra-violet.

F: slit of adjustable dimensions.

P: glass prism (crown or light flint) of 60° for visible and infra-red. Quartz prism for ultra-violet. A prism of rock-salt is more suited to work in infra-red.

In the visible region, it is often convenient to use a direct-vision prism (fig. 1.6) which consists of two prisms of crown *C* and one prism of flint *F* used in such a manner that the mean yellow light remains undeviated.

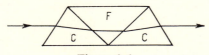

Figure 1.6

2*

The prism of Broca (fig. 1.7) is equivalent to a prism of 60° (*ABH* + *CAD*) and a total reflection prism *BHC*: it can deviate any desired radiation through 90°. One can make it in the form of a tank by cementing glass pieces with sugared arabic gum and filling it with carbon bisulphide, a very dispersive, liquid transparent in the near infra-red but not in the ultra-violet*.

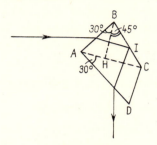

Figure 1.7

L: glass lens for visible and infra-red, quartz lens for ultra-violet. Focal length: 30 cm.

M: concave mirror with a radius of curvature of 60 cm.

R: grating.

E: white screen for visible; to detect ultra-violet a fluorescent screen is used (cf. § 22.6); to detect infra-red one can use a thermo-electric pile connected to a galvanometer.

b) Adjustments

Distance between S and F: for visible and infra-red a glass condenser, placed between *S* and *F*, forms an image of *S* on *F*.

For ultra-violet *F* is placed as near to *S* as possible.

Distance of F from L or from M: *L* or *M* form an image of *F* on a screen situated a few metres away.

Adjustment of F: its height is adjusted such that the light beam does not exceed the height of *P* or of *M*. Its width is adjusted empirically to obtain a compromise between the light intensity in the spectrum and its purity.

Adjustment of P: the setting-up of the prism for the minimum deviation position is not indispensable since the dispersion increases for large angles of emergence (but then astigmatism appears).

---

* Carbon bisulphide is highly inflammable. Ethyl cinnamate † is a safer but more expensive alternative. (The chemicals noted with a dagger are listed in Appendix A).

Adjustment of M: the grating is rotated so that the first order spectrum falls on the screen with maximum of light.

Adjustment of E: the screen is inclined about an axis parallel to *F* to obtain the spectrum sharply focused over all its extension.

For the study of ultra-violet and infra-red, half of the spectrum (half in height) should fall on a white screen and the other half on a special screen.

If a thermo-electric pile is employed, a slit of 1 to 2 mm in width cut in a tin foil is placed at the face of entry of the pile. This slit is displaced along a direction shown bythe trace of *E* in figures 1.4 and 1.5.

1.4 NEWTON'S EXPERIMENT

1.4.1 Principle

The superimposition of different radiations of the spectrum gives the sensation of white. This can be shown by the classical experiment using Newton's disc (fig. 1.8). The different sectors *A*, *B*, *C*, *D*, etc. ... have the colours of the spectrum, red, orange, yellow, green, etc. ... By rotating the disc rapidly, the colours look superposed due to the persistance of vision, and the disc appears white.

The experiment can equally well be carried out by the following arrangement (fig. 1.9). Carbon arc is at the focus of the lens O_1. Parallel rays

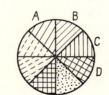

Figure 1.8

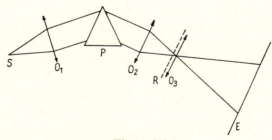

Figure 1.9

emerging from O_1 traverse the prism P and are dispersed. A spectrum is obtained in the focal plane O_3 of O_2. A lens placed in the plane O_3 projects the image of O_2 on a screen E. This image is white. This remains white even when a wire grating is placed on the spectrum.

1.4.2   Setting up the experiment (fig. 1.10)

a) Apparatus

　　S: carbon arc.
　　C: ordinary condenser.
　　F: vertical slit.
　　P: direct vision prism.
　　D: circular aperture.
　　L: achromatic lens of 20 to 30 cm focal length.
　　E: white screen.

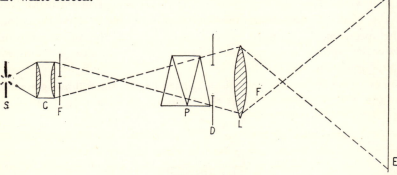

Figure 1.10

b) Adjustments

　　Adjustment of S, C and F: the distance between S and C is arbitrary. The slit F is placed against the condenser.

　　Adjustment of P, D, L and E: the direct vision prism is placed in the light beam in a manner that its exit face is well covered. The diaphragm D determines the limits of the beam. The lens L forms an image D' of D on the screen E.

　　The spectrum of S is observed at F', the image of F given by L. By masking certain regions of the spectrum F' by means of a screen, the image D' appears uniformly colored.

2

Interference Phenomena with Non-localized Fringes

2.1 PRELIMINARIES

2.1.1 Conditions for the vibrations to be able to interfere

In order that two light vibrations could interfere, they must:

— be coherent, that is, have well defined and constant phase relations between them.
— have the same frequency
— have parallel components.

With the exception of laser sources, interference can be observed only when the two vibrations come from the same source. Therefore, the wave coming from a source must be divided into two waves capable of interfering. There are two general methods:

1) interference by wave-front division
2) interference by amplitude division.

2.1.2 Interference by wavefront division

The following arrangements can be employed:

— Young's slits
— Fresnel's mirrors
— Fresnel's biprism
— Billet's split lens.

The figure 2.1 represents the scheme of Young's experiment. The two holes S_1 and S_2 of the screen E_1 isolate two different regions of the wave-

11

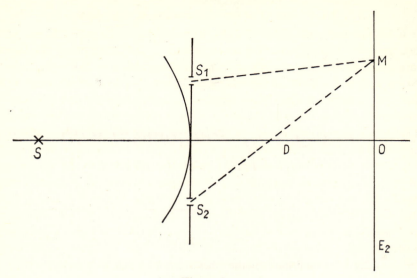

Figure 2.1

front Σ originating from a point source S. The two holes S_1 and S_2 act as two sources in phase if $SS_1 = SS_2$.

Let M be a point of the screen E_2 situated at a distance D from S_1S_2. Put $OM = y$ and $S_1S_2 = 2a$. The path difference between S_1M and S_2M is:

$$\delta = S_2M - S_1M = 2ay/D \qquad (2.1)$$

The interference fringes are approximately straight, parallel, equidistant and perpendicular to the plane of the figure. Bright fringes are obtained for $\delta = K\lambda$, K being a whole number. Dark fringes are obtained for

$$\delta = (2K + 1)\,\lambda/2$$

The central fringe at O is bright and the fringe spacing (distance between two consecutive dark or bright fringes) is: $i = \lambda D/2a$.

2.1.3   Interference by division of amplitude

In the interferometric arrangements of Michelson, Mach–Zehnder, Jamin etc. ... the incident wavefront is divided into two by means of a surface which is partially reflecting and partially transparent. Each ray gives rise to two rays. The fringes will be localized or non-localized depending upon the dimension of the source. If it is a point source, the fringes are not localized; if it is an extended source, the fringes are localized.

2.1.4   Aspect of the phenomenon in white light

It is known that radiations of different wavelengths do not interfere with one another. The illumination on the screen is thus a result of the super-imposition of various systems of fringes corresponding to different colours. At the centre, $\delta = 0$ for all the radiations. They all give a bright fringe in O and their superimposition gives again white light. As the point of observation is moved away from O, the phase difference depends on the wavelength. The two fringes on either side of the central white fringe appear dark. The next bright fringe is iridescent: bordered by violet on the inner side and by red on the outer (the wavelength of red being greater than the wavelength of violet), it appears more intense at the point where the path difference equals the wavelength for the yellow. This is due to the fact that it is in this region of the spectrum that the eye is most sensitive. When δ increases, at a particular point of the field certain wavelengths will give a minimum whereas some others will have a maximum. The resultant is the white colour called "white of the higher order". If a slit of a spectroscope is placed parallel to the fringes, at a point M of the field where we have white of higher order, a continuous spectrum will be observed crossed with dark lines (channelled spectrum) corresponding to well determined wavelengths. These lines correspond to those wavelengths which give a dark fringe at M.

2.2   YOUNG'S FRINGES

2.2.1   Principle

Consider two identical slits S_1 and S_2 made in an opaque screen. These slits are parallel to the slit source S and are equidistant from it (fig. 2.2). In accordance with Huygens principle, S_1 and S_2 can be considered to be two light sources. Since these are illuminated by the same source S, all the

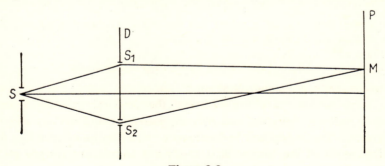

Figure 2.2

conditions for the vibrations emitted by S_1 and S_2 to interfere are fulfilled. The interference fringes are straight lines parallel to S, equidistant and perpendicular to the plane of the figure.

a) If the distance between S_1 and S_2 is altered, it is observed that the fringe spacing is modified. When the distance S_1S_2 is increased, the fringes get closer and when S_1S_2 is decreased, the fringes get farther apart.

b) Let us displace the slit S in its own plane, parallel to itself: the whole of the interference pattern gets displaced in the opposite direction.

c) If the slit S is rotated in its plane, interference fringes are rotated in the same direction.

d) On increasing the width of the slit S the visibility of the fringes decreases.

2.2.2 Setting up the experiment (fig. 2.3)

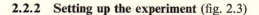
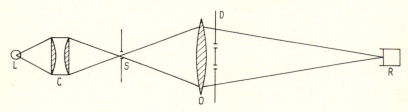

Figure 2.3

a) *Apparatus*

L: mercury vapour lamp of A_3 type with a green filter isolating 5461 Å line.

C: ordinary condenser of suitable diameter (ex. diameter of 10 cm and focal length of 5 to 7 cm) to collect a large portion of the flux emitted by L.

S: vertical slit of adjustable width.

O: lens of 35 cm focal length.

D: opaque screen, placed before the lens 0, with two narrow identical slits separated by a distance of about 5 mm.

R: television camera Vidicon without the objective, connected to the television receivers on which interference pattern is observed.

Filters: it may be necessary to introduce absorbing filters in the path of the light beam so that the photo-cathode of the camera does not get saturated.

b) Adjustments

Distance between L and C: arbitrary.

Distance between O and R: the lens O forms an image of S on the photo-cathode of the camera. The light path from O to R should be encircled with an opaque tube so that diffused light does not fall on the photo-cathode.

Distance between C and S: this distance is such that the condenser C forms an image of the source on S.

Distance between S and D: arbitrary.

2.2.3   Individual observation

A lamp with a straight vertical filament in front of which coloured filters can be held is placed on a high table. Opaque screens with slits are distributed to the students. Each screen carries three or four pairs of double-slits having different separations (of the order of 0.5 mm to 2 mm). The students can observe the influence of the separation of the slits on the fringe spacing. By using a red and a blue filter juxtaposed along a line normal to the length of the filament, it is observed that the red fringes are farther apart than the blue fringes.

The slits are made by scraping gelatin from over-exposed and developed photographic plates. Parallel slits can be obtained by placing two razor-blades on either side of a piece of cardboard. Alternatively, one can use two needles fixed in a cork.

2.3   FRESNEL'S MIRRORS

2.3.1   Principle

A source S illuminates two mirrors M_1 and M_2 forming between them an angle $\pi - \alpha$ (α being small). Interference fringes are observed in the region common to the two beams reflected from M_1 and M_2 (fig. 2.4). The two beams can be assumed to be coming from S_1 and S_2, the images of S given by M_1 and M_2. S_1 and S_2 are two synchronous and coherent sources since they are derived from the same source S. The fringes observed are approximately straight lines perpendicular to the plane of incidence and parallel to the line of intersection of the two mirrors.

2.3.2.   Setting up the experiment (fig. 2.5.)

a) Apparatus

L: carbon arc.

C: ordinary condenser.

S and T: vertical slits.

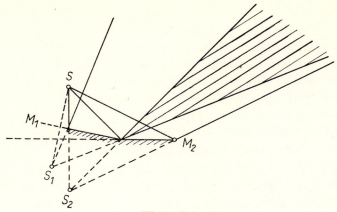

Figure 2.4

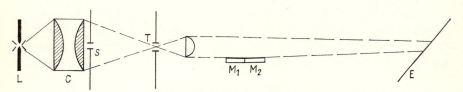

Figure 2.5

Interferometric arrangement: it consists of a plane-cylindric lens O of 2cm radius of curvature and of two mirrors M_1 and M_2. These are two pieces of plate glass whose back faces are silvered (or simply two pieces of black glass) to avoid parasite reflections. The two mirrors have a common edge.
E: white screen.

b) *Adjustments*

Distance between L and C: the distance is such that the image of L given by C is located at a distance of about 30 cm from C.

Distance between C and S: S is against the condenser.

Distance between C and T: the image of *L* formed by *C* is on *T*.

Distance between T and O: T is on the object focus of *O*.

Distance between S and the two mirrors: the cylindrical lens enlarges the light beam in a single direction increasing its brightness. By means of a screw the two mirrors are so adjusted that they have a well defined common edge and that the fringes be parallel to this edge. It is preferable to start this adjustment by employing a point source in place of the slit *S*.

Distance between the two mirrors and E: 7 to 8 metres. The screen should be inclined with respect to the light beams to have a maximum of fringe spacing. To start with, it is preferable to place the screen at a distance of 1 metre from M_1 and M_2. When sharp fringes are obtained, the screen is moved further away.

3

Interference Phenomena with Localized Fringes

3.1 CIRCULAR FRINGES OF EQUAL INCLINATION TO A PLANE PARALLEL PLATE

3.1.1 Principle

Consider a plane parallel plate of thickness e and of refractive index n. S is a monochromatic extended source. The incident ray SI_1 (fig. 3.1) which has an angle of incidence i and an angle of refraction r is broken up into

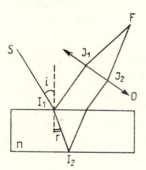

Figure 3.1

two rays by the plate which is at the same time reflecting and transmitting. These two rays I_1J_1 and I_2J_2, interfere at infinity or in the focal plane of a lens O. At infinity, the path difference between these two rays is:

$$\delta = 2ne \cos r + \lambda/2 \qquad (3.1)$$

The term $\lambda/2$ is due to the reflection air-plate of the ray SI_1. The formula (3.1) shows that δ is constant if r is constant. The fringes are thus circles centered on a normal to the plate. These are circles of equal inclination localised at infinity.

The same phenomenon is observed in transmission (the term $\lambda/2$ disappears from the formula (3.1)). But the contrast is less good and it is better to observe the phenomenon by reflected light.

3.1.2   Setting up the experiment (fig. 3.2)

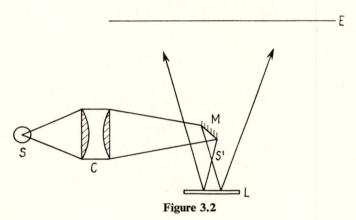

Figure 3.2

a) Apparatus

   S: mercury vapour lamp of medium pressure (A_2 type) provided with a filter isolating the green line.

   C: ordinary condenser of large aperture (diameter of about 15 cm and focal length of 5 to 6 cm).

   L: sheet of mica with parallel faces and a thickness of about 5/100 mm.

   M: mirror, of as small dimensions as possible, which reflects the light coming from the condenser on to the sheet L.

   E: white screen.

b) Adjustments

   Distance between S and C: about 15 cm; to avoid parasite light a black tube should be placed between the lamp and the condenser.

   Distance between C and M: about 10 cm. The mirror which reflects the light beam coming from the condenser in a direction perpendicular to the axis SC, must be placed in front of S', the image of S given by C.

Distance between M and L: L is slightly beyond the point of convergence of rays coming from *M* or it is slightly before this point. This adjustment is done by trial and error such that the shadow of the mirror *M* on the screen *E* occupies only a small region in the centre of the circular fringes. Only a small area of the sheet *L* should be used in order to ensure uniformity of thickness.

Distance between L and E: E is placed at about 4 to 5 meters from *L*. Under these conditions it is practically at infinity. Besides, the source *S* is not very large and therefore the fringes are not perfectly lokalized at infinity.

3.1.3   Individual observation

The stage of the lecture theatre (and particularly the white screen) is illuminated by a strong sodium lamp (A_1 type). Each student is given a sheet of cleaved mica; he places it against one of his eyes and looks through it on to an illuminated surface. He perceives circular fringes of low contrast. These are the fringes of equal inclination as seen by transmission. The converging system of the eye, which is unaccommodated, plays the role of *O* (fig. 3.1); *F* is on the retina.

3.2   FRINGES OF THIN FILMS

3.2.1   Principle

a) Plate of variable thickness in monochromatic light

A thin plate *L* of slowly varying thickness *e* and of constant refractive index *n* is illuminated by a beam of monochromatic light coming from a source *S* at infinity. Figure 3.3 represents two parallel rays coming from a single point of the source *S*. The incidence is nearly normal. The ray *SI*

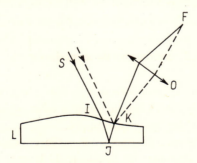

Figure 3.3

is refracted at I and again at K after suffering reflection at J, from the back face of the plate. The ray SK, parallel to SI and originating from the same point of the source, gets reflected at K. The two rays meet at F, the image of K given by the lens O. In practice, the points I, J and K are extremely close to one another and it can be said that the lens O forms an image of the thin plate L at F. The rays $SIJKF$ and SKF are capable of interfering and the interference phenomenon, localized practically on the thin film, is observable on the screen placed at F. Since the incidence is close to normal, the points I and K are practically overlapping and the thickness of the plate is the same at these two points.

The path difference between these two rays is:

$$\delta = 2ne + \lambda/2 \qquad (3.2)$$

The interference fringes are the lines $ne = $ constant; these are the lines of equal optical thickness of the plate. This phenomenon can be observed with thin films which do not have rigorously parallel faces, such as a film of soapy water and the air film between touching glass plates.

b) Phenomena in white light

Thin films with thickness not too large, $ne = 0.2$ to 1 micron, show bright colours when illuminated in white light. Each component radiation gives a system of fringes and at a particular point of the film the order of inter- ference $p = \delta/\lambda$ varies with wavelength. The radiations for which $p = K + \frac{1}{2}$ (K is a whole number) are not reflected and the colours of the film at this point results from the combination of remaining radiations. Since the thickness of the film is not constant, the observed colours are not the same at every point.

c) Observation

The fringes exist in reflected light as well as in transmitted light but, as in § 3.2.1, the observation is made in reflected light because the contrast is better.

3.2.2   Setting-up the experiment (fig. 3.4)

a) Apparatus

S: for observation in monochromatic light: mercury vapour lamp of medium pressure (type A_2) provided with a filter which isolates green line. For observation in white light: carbon arc.

C: ordinary condenser of dimensions large enough to collect the light issuing from S (ex: diameter of 10 cm, focal length of 5 to 7 cm).

M: plane mirror

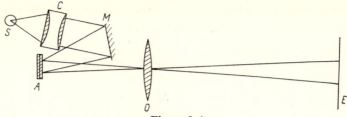

Figure 3.4

A: the film to be studied is an air film included between two glass plates. Since the glass surfaces are not perfectly plane, the air film has a variable thickness. The glass plates are held against each other in a metal frame. With the help of three screws the thickness of the air film can be varied. The outer faces of the two glass plates make a certain angle (5 to 6 degrees) with the inner faces. This permits the elimination of parasite reflections (fig. 3.5). The film *A* can equally well be a glass plate of varying thickness, a plane film of soapy water etc....

O: lens of about 30 cm focal length and of a diameter sufficient to collect all the light reflected from *A*.

E: white screen.

Figure 3.5

b) Adjustments

Distance between S and C: the source *S* is very near to the object focus of *C*. The beam coming from *C* must give a real image of *S* at a distance of say 80 cm.

Distances between C and M and between M and A: arbitrary, provided that the image of *S* is formed on *O* after reflection of the light beam from *M* and *A*. The mirror *M* should be so oriented that the reflected beam falls on *A* at an incidence close to normal.

Distance between A and O: *A* is very near to the object focus of *O*.

Distance between O and E: *O* forms an image *A'* of *A* at a distance of about 4 m on the screen *E*.

3*

3.2.3   Simultaneous observation of the phenomena by transmission and by reflection (fig. 3.6)

The scheme is that of figure 3.4 for the observation by reflection. The light beam transmitted by the plate A falls on a mirror M'. A lens O' of 30 cm focal length is placed at S'', image of the source given by the condenser after reflection of the light beam from M and M'. This lens forms an image of A on the screen E after reflection of the light beam from M'. The orientation and the position of M' is adjusted in order to obtain the two images of A side by side on the screen E.

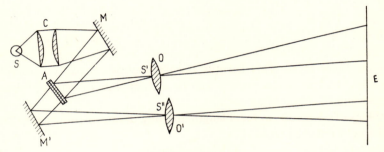

Figure 3.6

3.2.4.   Individual observation

The observer adjusts on the fragments of thin films $L, L', \ldots$ (fig. 3.7) placed on a black background N of paper (or better of velvet) spread over a table. The source S is a sodium lamp (type A_1) placed behind a large ground glass V. The films are fragments of cleaved mica or microscope slide-covers. Thin

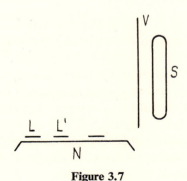

Figure 3.7

films of glass can be obtained by blowing into a glass tube, with one end closed, softened in a flame, until it breaks down. These phenomena can also be observed in an air film contained between two fragments of glass placed one over the other. The fringes are localized on the thin films.

3.3 AIR WEDGE

3.3.1 Principle

Consider a thin film of air contained between two glass plates at a small angle to each other (fig. 3.8). The two surfaces of the glass plates which face each other are plane. The interference fringes (that is the loci of equal thickness of the air film) are equidistant straight lines parallel to the line of intersection of the glass plates and having a fringe spacing of $\lambda/2\theta$.

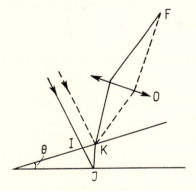

Figure 3.8

3.3.2 Setting up the experiment

The scheme is that of figure 3.4 in which the element A is replaced by two glass plates with their plane faces making an angle θ with each other.

3.4 NEWTON'S RINGS

3.4.1 Principle

The interference fringes obtained from an air film contained between a spherical and a plane surface (fig. 3.9) are circular in form. If the spherical surface is in contact with the plane surface, the central circle is dark. In this case, the radii of dark circles are proportional to the square roots of successive whole numbers.

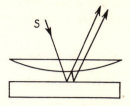

Figure 3.9

3.4.2 Setting up the experiment

The scheme is that of figure 3.4. The element A is replaced by the system represented in figure 3.9. The radius of curvature of the plano-convex lens is of the order of 100 metres.

3.5 CHANNELED SPECTRUM GIVEN BY THIN PLATES

3.5.1 Principle

The light reflected from a transparent plate which is too thick to give interference colours is white of higher order. Spectral analysis of the reflected light shows dark bands for the radiations λ_1, λ_2 etc. ... satisfying the relation

$$2ne + \lambda/2 = (2K + 1)\lambda/2 \qquad (3.3)$$

(K is a whole number)
or:

$$2e = K_1\lambda_1/n_1 = K_2\lambda_2/n_2 = \ldots \qquad (3.4)$$

In these two expressions the dispersion of refractive index is neglected.

3.5.2 Setting up the experiment (fig. 3.10)

a) *Apparatus*

  S: carbon arc.

  C: ordinary condenser.

  F: vertical slit of adjustable width.

  O_1: lens of about 35 cm focal length and a diameter sufficient to collect all the light diffracted by F.

  A: thin film to be studied (in this case, it is the air wedge of § 3.2.2).

O_2: lens with a focal length of about 30 cm and with a diameter sufficient to collect all the light reflected by A.

R: a plane reflection grating of good quality.

E: white screen.

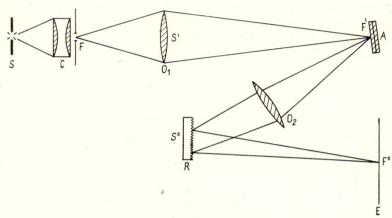

Figure 3.10

b) Adjustments

Distance between S and C: this distance is such that the image S' of S given by C should be at a distance of about 65 cm from C.

Position of F: F is placed against C.

Distance between F and O_1: the lens O_1 is placed at S'.

Distance between O_1 and A: this distance is such that O_1 forms an image F' of F on A. The image F' must be parallel to the fringes of the air wedge. The air wedge A is so disposed that the light issuing from O_1 is incident on it in near normal direction.

Distance between A and O_2: about 30 cm; this distance is so chosen that O_2 forms an image F'' of F at a distance of about 4 m and that the image S'' of S' be situated at about 20 cm from O_2 after the reflection A.

Distance between O_2 and R: the grating R is placed at S'' and covers completely the image of the arc. Thus all the light issuing from the arc is concentrated on the grating.

Distance between R and E: about 4 m. The image F'' of F given by O_2 is formed on E via the grating R.

3.5.3 Alternative set-up (fig. 3.11)

a) *Apparatus*

  S: carbon arc.
  C: ordinary condenser.
  M; cleaved sheet of mica having a thickness of the order of 5 to 10 μ.
  A: thick rod of a carbon arc.
  L: lens with a focal length of 15 to 20 cm.
  P: direct vision prism.
  E: white screen.

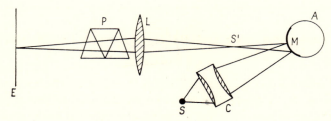

Figure 3.11

b) *Adjustments*

   Distance between S and C: S is very near to the object focus of C. The
light beam coming from C is very slightly converging. It illuminates the thin
(a few microns) mica sheet M which is rolled on the carbon rod A placed
in a vertical position.
   Adjustment of L, M and E: the mica sheet M functions as a cylindrical
mirror giving, in its vicinity, a vertical straight image S' of S. The lens L
forms on E an image of S'.
   Adjustment of P: the introduction of P with its edge in a vertical position
gives a channeled spectrum. The thinner the mica sheet, the lesser is the
number of bands in the channeled spectrum.

4

Michelson Interferometer

4.1 PRINCIPLE AND ADJUSTMENT OF THE INSTRUMENT

4.1.1 Principle

The point source S is at the focus of the lens O_1 (fig. 4.1). The beam of parallel rays given by O_1 is broken up into two by the semi-reflecting plate G_1, called the beam splitter. One half of the beam is reflected by the plane mirror M_1 and the other half by the plane mirror M_2. The two beams recombine, one having been reflected by G_1 and the other having traversed G_1, and interfere. It may be noted that the beam reflected by M_2 has passed through G_1 three times, whereas the other beam has traversed it only once. To compensate for this path difference a plane parallel plate G_2, the com-

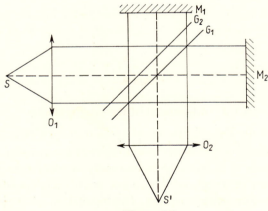

Figure 4.1

29

pensator, made of the same glass and of same thickness as G_1, is introduced. in the path corresponding to the mirror M_1. The plate G_2 is parallel to G_1. Thus when the two mirrors are symmetrical with respect to the beam-splitter, the beams SAM_1AO_2 and SAM_2AO_2 have traversed equal distances and their path difference is zero.

M_2' is an image of the mirror M_2 given by the reflecting surface of the beam splitter. The interference pattern observed can be considered to be produced by a film of air contained between M_1 and M_2' (fig. 4.2).

If M_1 and M_2' are rigorously parallel, circular fringes of the air film are observed at infinity by replacing the point source and the lens O_1 by an extended source. These circular fringes are localised in the focal plane of the lens O_2.

If M_1 and M_2' are not parallel and make a small angle between them, fringes of the air wedge localised on M_1 are observed with an extended source.

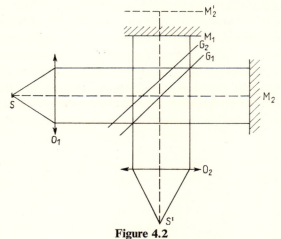

Figure 4.2

4.1.2   Adjustment of the interferometer

a) Adjustment for the observation of circular fringes at infinity

Setting the beam splitter and the compensating plate parallel to each other: The interferometer is illuminated by normally incident light using a point source. A number of images of this source are observed through the beam splitter and the compensating plate. The inclination of the compensating plate is adjusted by means of a screw so as to superpose the various images.

Setting M_1 and M_2' roughly parallel to each other: The point source is placed in the focal plane of O_1. Different images of this source are given by the two mirrors. By means of the fast motion screw the inclination of the mirrors is adjusted so that these images are superposed. Thus M_1 and M_2' are roughly parallel to each other.

Setting M_1 and M_2' accurately parallel to each other: The interferometer is illuminated by an extended source (for example a mercury vapour lamp of medium pressure of A_3 type). To render the illumination uniform, a ground glass screen is placed in front of the lamp. The circular fringes at infinity are thus observed. To render them perfectly circular accurate adjustment for parallelism is made by means of slow motion screws provided on the back of the mirrors.

a) Adjustment for the observation of fringes of the air-wedge.

Having adjusted the interferometer for circular fringes at infinity, it suffices to incline one mirror with respect to the other. The straight line fringes of the air wedge are thus observable.

4.2 OBSERVATION OF THE FRINGES OF EQUAL INCLINATION OF AN AIR FILM BY MEANS OF THE MICHELSON INTERFEROMETER

4.2.1 Principle

The mirror M_1 and the image M_2' of the mirror M_2 are parallel (fig. 4.2). If e is the distance $M_1 M_2'$, the path difference between the two interfering rays is:

$$\delta = 2ne \cos r + \lambda/2 \qquad (4.1)$$

Here $n = 1$ since we are using an air film.

Varying the thickness of the film, that is, by displacing the mirror M_2 along its surface normal, the diameter of the rings can be varied. If e decreases, the rings disappear at the centre and their radii increase. If e increases, the rings originate at the centre and their radii decrease. There are more and more rings in the field.

When the image M_2' of M_2 coincides with M_1, the rings disappear and the field appears uniformly coloured. The interferometer is said to be set for a path difference of zero.

If the beam splitter and the compensating plate are accurately parallel, the rings at infinity are circular. When one is inclined with respect to the other, they get deformed. They become ellipses and if the inclination is increased they change to hyperbolas.

4.2.2   Setting up the experiment (fig. 4.3)

a) Apparatus

S: mercury vapour lamp of medium pressure of type A_3 with a filter isolating the green line.

C: ordinary condenser of large aperture (diameter $D = 15$ cm and the focal length $f = 5$ to 6 cm).

M: Michelson interferometer is adjusted for the observation of rings at infinity.

E: white screen.

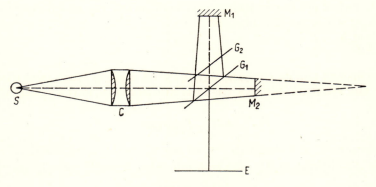

Figure 4.3

b) Adjustments

Distance between S and C: 25 cm approximately.

Distance between C and M: about 5 cm. This distance is such that the condenser forms an image of the source S slightly beyond the mirror M_2 so as to illuminate only the reflecting surface of the mirror. All the light emitted by S should arrive on the screen E.

Distance between M and E: 3 to 4 metres. The screen E is, to a good approximation, at infinity with respect to the air film $M_1 M_2'$.

4.3 OBSERVATION OF THE FRINGES OF AN AIR-WEDGE WITH A MICHELSON INTERFEROMETER

4.3.1 Principle

a) Observation in monochromatic light

The mirror M_1 and the image M_2' of the mirror M_2 form between them a small angle θ (fig. 4.4). With an extended source S and under quasinormal incidence, straight line, parallel and equidistant fringes are observed. The fringe spacing is $\lambda/2\,\theta$ and they are localised on the mirror M_1. By modifying the inclination of one of the mirrors (thus the angle of the air wedge) the spacing and the number of the fringes can be varied.

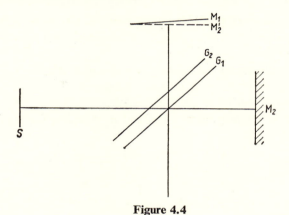

Figure 4.4

On decreasing the diameter of the source, the contrast of the fringes increases though there is less light in the interference pattern. On sufficiently decreasing the source diameter, non-localized fringes are observed. When the incidence is varied from near normal, the fringe contrast diminishes rapidly. On varying the value of δ by heating the air in the vicinity of the mirrors (this modifies the refractive index of the air) a deformation of the fringes is observed.

b) In polychromatic light

If the source consists of a number of spectral lines, the contrast of fringes varies rapidly as δ increases (cf. 4.4).

In white light, fringes are seen only when the thickness of the air wedge is almost zero. The interferometer must, therefore, be adjusted for a path-difference of zero, the mirrors being parallel, and then one of the mirrors must be inclined with respect to the other. Thus two or three iridescent fringes are observed on either side of the central fringe and then there is the white of the higher order.

c) *Channelled spectrum*

If a slit, parallel to the fringes, is interposed in the white of the higher order followed by a direct vision prism (or a grating) and a projection lens a channelled spectrum is observed on the screen.

4.3.2 Observation of the fringes of an air-wedge (fig. 4.5)

a) *Apparatus*

S: mercury vapour lamp of medium pressure (A_3 type) with a filter for observation in monochromatic light or a carbon arc for observation in white light.

C: ordinary condenser with a diameter of about 10 cm.

D: diaphragm with holes of different diameters.

L_1: achromatic lens of 25 cm focal length.

M: Michelson interferometer adjusted for observing the fringes of an air wedge.

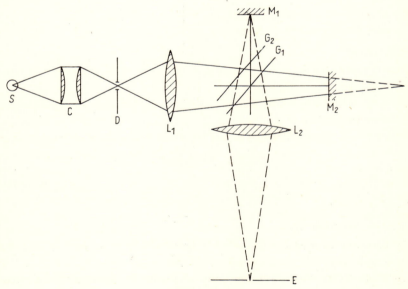

Figure 4.5

L_2: achromatic lens of 15 cm focal length.

E: white screen.

b) *Adjustments*

Distance between S and C: about 10 cm.

Distance between C and D: D is placed at S', the image of S given by C.

Distance between D and L_1: L_1 is placed at about 30 cm from D and gives an image S'' of S' some distance away. The beam of light must not make a large angle with the axis of the system.

Distance between L_1 and M: M is placed at such a distance from L_1 that the light beam coming from L_1 completely covers the mirror M_2.

Distance between M and L_2: L_2 projects the mirror M_1 on the screen E. M_1 is placed very near to the focus of L_2.

Distance between M and E: 4 to 5 m.

By making D a few millimeters in diameter, L_2 may be removed and E may be placed at any distance without causing the fringes to disappear.

4.3.3.   Observation of a channelled spectrum (fig. 4.6)

a) *Apparatus*

The same material as in 4.3.2 and in addition

F: slit.

P: direct vision prism.

L_3: lens of 20 cm focal length.

b) *Adjustments*

Distance between M and L_2: L_2 is at distance of 30 cm from M_1 and gives an image M_1 at 30 cm.

Distance between L_2 and F: F is in the plane of M_1' and is oriented parallel to the fringes.

Position of P, L_3 and E: L_3 gives an image of F, through P, on the screen E.

4.3.4   Variation

The experiment illustrated in figure 3.9 is set up. The slit F is projected on the mirror M_2 of the interferometer and the light reflected from the mirror M_1 is analysed with the grating R.

4.4   OBSERVATON OF BEATS

4.4.1   Principle

An interferometer permitting a continued variation of the path difference is illuminated by a source emitting two monochromatic radiations of wave-

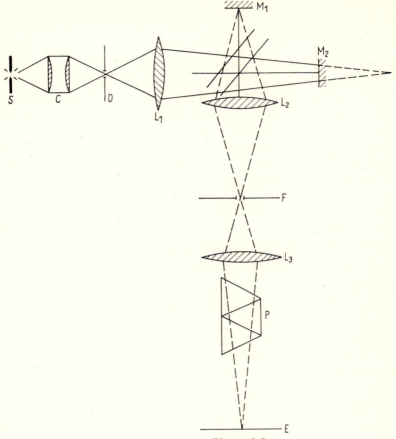

Figure 4.6

lengths λ and λ' and of nearly equal intensities. Each radiation gives a system of fringes. In order that the fringes be sharp, the minima and maxima corresponding to one radiation should superpose respectively on the minima and maxima corresponding to the other radiation. For the maxima, we have:

$$\delta = p\lambda = (p + m)\lambda' \qquad \lambda > \lambda' \tag{4.2}$$

p and m being whole numbers.

With $p = 0$, $\delta = m = 0$, the fringes are sharp. If δ increases, we arrive at $m = 0{,}5$: a dark fringe of one of the systems coincides with a bright fringe of the other and the fringes get diffused. For $m = 1$, the fringes again become sharp. The variation $\varDelta\delta$ of δ between two consecutive co-incidences is such that:

$$\varDelta\delta = p\lambda = (p + 1)\,\lambda' \qquad (4.3)$$

whence:

$$\varDelta\delta = \frac{\lambda\lambda'}{\lambda - \lambda'} \qquad (4.4)$$

$\varDelta\delta$ is the period of coincidences.

A Michelson interferometer set for the observation of rings at infinity and with a path difference of zero to start with, is used for the experiment. If one of the mirrors is displaced through $\varDelta e$ along its normal, we have $\varDelta\delta = 2\varDelta e$.

4.4.2 Setting up the experiment (fig. 4.7)

a) Apparatus

S: sodium vapour or mercury vapor lamp with a yellow filter transmitting the yellow and green lines. The phenomenon is much less sharp than that observed with the two radiations of sodium though the period of coinci-dences is small and one can see in the field of view a number of concordances and discordances at the same time without the fringes being too narrow.

C: condenser with a focal length of 5 to 7 cm.

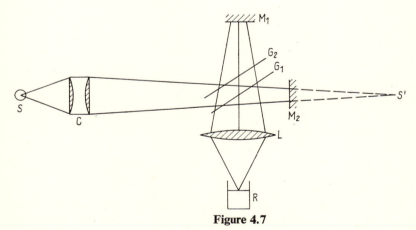

Figure 4.7

M: Michelson interferometer adjusted for the observation of rings at infinity.

L: lens with a focal length of about 15 cm

R: television camera, vidicon, without objective, related to the television receiver on which the phenomenon is observed.

b) *Adjustments*

Distance between S and C: about 25 cm.

Distance between C and M: about 5 cm. The light beam coming out of C must cover exactly the mirror M_2.

Distance between M and L: arbitrary.

Distance between L and R: the photo-cathode of the camera is placed in the image focal plane of the lens *L.*

5

Temporal Coherence and Spatial Coherence

5.1 DEFINITIONS

5.1.1 Temporal Coherence

A source of light, even if it is a point, consists of a large number of atoms which emit wave-trains. For simplicity, it is assumed that all the atoms emit wave-trains of the same duration τ, called the coherence-time.

If τ is infinite, the light emitted by the source is monochromatic; this is never the case in practice. The thermal sources of highest monochromaticity have a coherence time τ of the order of 0.6×10^{-9} seconds for $\lambda = 0.5\,\mu$.

Let us consider a Michelson interferometer illuminated by a point source S (fig. 5.1). At an instant t a wave train T is broken up by G into two wave trains T' and T''. If the path HM' followed by the wave train T' is shorter than the path HM'' followed by the wave train T'', the wave-train T'' arrives at O at a time 2θ after the arrival of T'. When 2θ is less than τ, the two wave-trains which superpose at O belong to a single incident wave-train. When 2θ is greater than τ, these wave trains belong to two different wave-trains which were emitted at different instants of time by the source S.

In the first case, the two wave-trains having originated from the same incident wave-train, the interference phenomenon in O does not depend on the instant of time at which this wave-train was emitted by the source and therefore a system of interference fringes is observed at O.

In the second case, the interference pattern in O varies with each wave train emitted by the source. Since a large number of wave-trains arrive at O during the interval of time necessary to make an observation, the

4*   39

superposition of different systems of fringes corresponding to different wavetrains will overwhelm the interference phenomenon and give rise to uniform intensity at O.

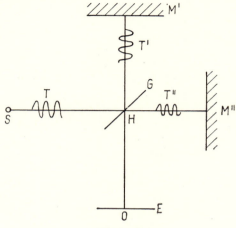

Figure 5.1

5.1.2 Degree of Coherence in quasi-monochromatic light

Two points P' and P'' of a plane E_1 (fig. 5.2) are illuminated by an ensemble of light vibrations emitted by an extended, incoherent and quasi-monochromatic source S.

According to the Van Cittert–Zernike theorem, the degree of spatial coherence of light vibrations at P' and P'' is given by the formula:

$$\Gamma(yz) = \frac{\iint\limits_{S} I(\beta\gamma)\, e^{jk(\beta y + \gamma z)} \mathrm{d}\beta\, \mathrm{d}\gamma}{\iint\limits_{S} I(\beta\gamma)\, \mathrm{d}\beta\, \mathrm{d}\gamma} \tag{5.1}$$

where $I(\beta, \gamma)$ is the distribution of energy on the source S. If two holes are made at the points P' and P'' in the screen E_1, Young's fringes can be observed on a second screen E_2. The visibility of Young's fringes on the screen E_2 will be given by the modulus of Γ. The visibility of Young's fringes can thus be foreseen as a function of the width of the source.

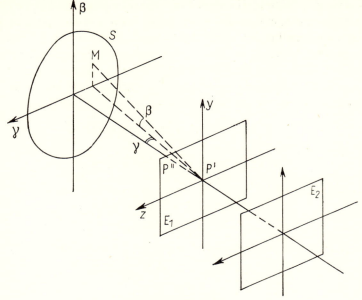

Figure 5.2

5.2 EXPERIMENT SHOWING THE VARIATION OF THE CONTRAST OF THE FRINGES OF EQUAL THICKNESS AS A FUNCTION OF THE PATH DIFFERENCE AND THE MONOCHROMATICITY OF THE SOURCE (TEMPORAL COHERENCE)

5.2.1. Principle

Michelson interferometer is adjusted for approximately zero path difference. The fringes of an air-wedge are observed on illuminating the interferometer by a laser (a point source).

It is demonstrated that on increasing the path difference the visibility of fringes remains excellent. The same experiment is repeated with another source, a mercury vapour lamp (green line) for example. The visibility of fringes decreases on increasing the path difference even if the source is a point.

The monochromaticity of a source is related to the coherence time τ. The smaller the vapour pressure of mercury in the lamp, the smaller is the band-width of the radiation emitted and the longer is the coherence time τ. This phenomenon can be observed with a mercury vapour lamp placed before a Michelson interferometer. The lamp is switched on and till it reaches its steady state, fringes of an air-wedge will be visible in high contrast though these are not very bright; as the light intensity emitted by the lamp rises, fringe contrast decreases (as the vapour pressure of mercury increases the intensity increases) and when the lamp attains its steady state, the fringes may even disappear if the path difference between the two arms of the Michelson is large (a few tenths of a millimeter with a lamp of the A_3 type).

5.2.2   Setting up the experiment

The apparatus is as shown in figure 4.5.

5.3   SAME EXPERIMENT REPEATED WITH CIRCULAR FRINGES IN THE FAR FIELD OF A MICHELSON INTERFEROMETER (TEMPORAL COHERENCE)

5.3.1   Principle

The same phenomena are observed as in the preceeding experiment.

5.3.2   Setting-up the experiment

See figures 4.3 and 4.7.

5.4   VARIATION OF THE VISIBILITY OF FRINGES OF EQUAL THICKNESS AND OF YOUNG'S FRINGES AS A FUNCTION OF THE DIMENSIONS OF A MONOCHROMATIC SOURCE (SPATIAL COHERENCE)

5.4.1   Principle

The Michelson interferometer is illuminated in monochromatic light and is adjusted for approximately zero path difference. The theorem of Van Cittert–Zernike enables us to visualize a decrease in the contrast of fringes when the source is enlarged. For a certain diameter of source the fringe contrast becomes zero, and if we continue to enlarge the source, fringes reappear. On reappearance the fringes have low contrast and the bright and dark fringes are observed to have interchanged their position. The contrast of the fringes continues to decrease; it becomes zero and the second reappearance of the fringes is difficult to observe since the contrast is lower.

Exactly identical phenomenon is observed with Young's fringes. If I_1 and I_2 are the intensities of the bright and dark fringes respectively, the visibility Γ of the fringes is defined by the ratio:

$$\Gamma = \frac{I_1 - I_2}{I_1 + I_2} \tag{5.2}$$

The curve of figure 5.3 shows the variations of Γ as a function of the width d of the slit S. After the value d_0 there is an inversion of the fringe visibility.

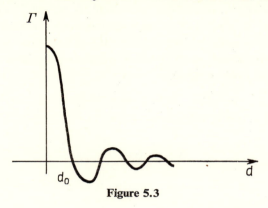

Figure 5.3

5.5 FOURIER SPECTROSCOPY

5.5.1 Principle

The Michelson interferometer is illuminated by a polychromatic source S and is adjusted for the observation of circular fringes at infinity. A photo-

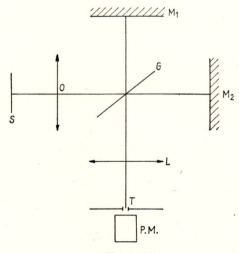

Figure 5.4

multiplier *PM* records the light intensity in the focal plane of the lens *L* (fig. 5.4). An opaque screen with a hole *T* is placed in the focal plane *F*. The hole *T* isolates a small region of uniform intensity at the centre of circular fringes. To a value Δ of the path difference between the two rays interfering at *T*, corresponds the intensity $I(\Delta)$. If $B_0(\sigma)$ is the brightness of the source *S* for the wavelength $\lambda = 1/\sigma$ (σ_1 and σ_2 are the limits of the spectral interval of the source) we have:

$$I(\Delta) = \int_{\sigma_1}^{\sigma_2} B_0(\sigma)\,\mathrm{d}\sigma + \int_{\sigma_1}^{\sigma_2} B_0(\sigma) \cos 2\pi\sigma\Delta\,\mathrm{d}\sigma \tag{5.3}$$

$$I(\Delta) = I_1 + I_0(\Delta) \tag{5.4}$$

$I(\Delta)$ is equal to the sum of a constant term I_1 and a variable term $I_0(\Delta)$. Δ is varied continuously by displacing the mirror M_1 (or M_2) and the photo-multiplier enables the tracing of the curve $I(\Delta)$ as a function of Δ (fig. 5.5).

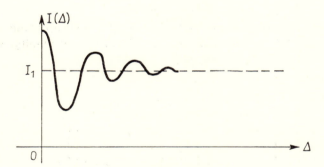

Figure 5.5

The curve $I_0(\Delta) = I(\Delta) - I_1$ is called the interferogram (fig. 5.6). The oscillations of the curve represent the variations of the visibility of fringes. They decrease as the path difference increases. The spectrum of the source *S* is obtained by calculating the Fourier transform of the interferogram.

In figure 5.7 we have represented (after Michelson) spectra of different sources and the corresponding interferograms.

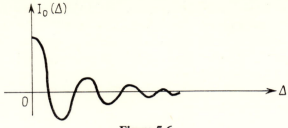

Figure 5.6

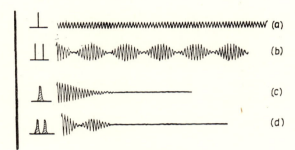

Figure 5.7

5.5.2 Setting-up the experiment (fig. 5.4)

a) Apparatus

S: light source under study.

L and L': two lenses of arbitrary focal lengths.

Michelson interferometer M: it is adjusted such that the mirrors M_1 and M_2 are parallel. A motor allows continuous movement of mirror M_2.

Recorder R: a photo-multiplier (RCA IP 28 for example) with power-supply and a recorder.

b) Adjustments

Adjustments of S and L and of R and L': S is at the object focus of L and the cathode of the photo-multiplier is at the image focus of L'.

Distance between M and L': arbitrary.

6

Multiple Beam Interference Produced by Semi-reflecting Plates

6.1   PRINCIPLE OF THE PHENOMENA OBSERVED IN TRANSMITTED LIGHT

Consider two semi-reflecting, plane and parallel surfaces, AB and $A'B'$ (fig. 6.1), bounding a medium of thickness AA' and of refractive index n. An incident ray SI gives rise to a series of reflected rays and a set of parallel

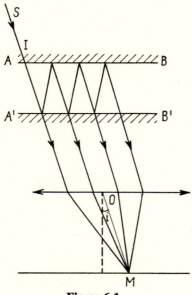

Figure 6.1

47

rays emerges. These rays interfere in the focal plane of the lens O. We suppose that the reflectivity and the transmissivity of the medium above AB are the same as those of the medium below $A'B'$.

Let T and R be the reflectivity and the transmissivity in terms of flux of the surface AB and $A'B'$. Let us put

$$I_0 = \frac{T^2}{(1 - R)^2} \qquad A = \frac{4R}{(1 - R)^2} \qquad (6.1)$$

and

$$\Phi = \frac{4ne \cos i}{\lambda} + 2\varphi' \qquad (6.2)$$

where φ' is the phase change due to reflection. The intensity at a point M of the focal plane of the lens O is given by the formula:

$$I_M = \frac{I_0}{1 + A \sin^2 \dfrac{\Phi}{2}} \qquad (6.3)$$

If a number of spectral lines illuminate simultaneously a multiple-beam interferometer, each line gives a system of bright rings in transmitted light. A superimposition of these systems will be observed. When the rings are sharp and fine, the different systems will juxtapose without getting confused. Such an interferometric instrument can, thus, work as a spectroscope.

The phenomena can equally well be observed in reflected light, but they are not complementary to the phenomena observed in transmitted light, except in the case when there is absolutely no absorption.

6.2 EXPERIMENT SHOWING THE SHARP RINGS: FABRY–PEROT INTERFEROMETER

a) Apparatus

  S: mercury vapour lamp of medium pressure, type A_3.

  C: condenser with a diameter of 10 cm and a focal length of 5 to 7 cm.

  Fabry–Perot interferometer: the instrument consists of two semi-silvered glass plates. Multiple-beam fringes of the air film contained between the two plates, are observed. One glass plate is fixed and the second is mounted on a moveable carriage. The two glass plates are made parallel by means of screws provided for the purpose. By displacing the moveable glass plate, the thickness of the air film can be regulated. The rails on which the moveable carriage rests are so constructed that the parallelism is not destroyed during the movement of the mirror.

For the usual experiments, there is no need to vary the thickness of the air film and it suffices to use an interferometric standard (fig. 6.3). The two semi-silvered mirrors are separated by three spacers of suitable thickness. It is very difficult to obtain three spacers of identical thickness. The parallelism is attained by forcing suitably the mirrors against the spacers C by

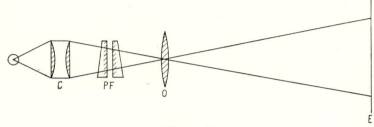

Figure 6.2

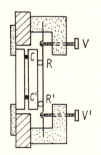

Figure 6.3

means of three springs R which are worked by three screws V. The mirrors are parallel when the diameter of rings no longer depends on the part of the air film used.

O: lens of large diameter and of long focal length (for example 2 m).

E: white screen placed in the focal plane of O.

b) Adjustments

Distance between S and C: the condenser gives an image S' which is approximately at the same distance from C as the source.

Distance between C and the Fabry–Perot: it is advantageous to put the Fabry–Perot near the lens O which is placed at S'. A tube must be placed between the condenser and the Fabry–Perot to avoid the parasite light. It is preferable to work with films of small thickness in the Fabry–Perot, for example, 0.5 mm.

Distance between O and E: if the mirrors of the Fabry–Perot have high reflectivity, that is, the fringes are sharp, it is necessary to make the observations in the focal plane of the lens *O*. In order to obtain sufficient light with a lamp of A_3 type, the focal length should not be more than 2 m.

6.3 EXPERIMENT WITH MIRRORS HAVING RELATIVELY LOW REFLECTIVITY (fig. 6.4)

a) Apparatus

S: mercury vapour lamp of medium pressure (A_3) provided with a filter isolating the green line.

C: condenser of large aperture, 15 cm in diameter and 5 to 7 cm of focal length.

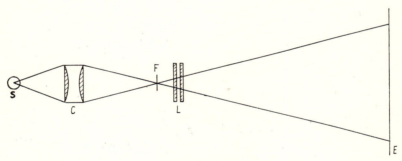

Figure 6.4

L: air film comprised between two glass plates the opposite faces of which are semi-reflecting and parallel. In this experiment, we are not interested in obtaining very sharp and fine fringes rather in obtaining lot of light. The semi-reflecting films on the glass plates are deposited with this in view; thus materials with as low absorbence as possible are used (thin dielectric films). If the air film is very thin (thickness of a cigarette paper) the rings are very large and sufficiently sharp on a screen placed at a distance of about 3 metres and there is no necessity to observe in the focal plane of a lens.

b) Adjustments

Distance between S and C: this is such that the image *S'* of the source is situated at about the same distance from *C* as the source.

Position of the filter and the film L: close to *S'*.

Distance between L and E: of the order of 3 meters.

6.4   USE OF A HE–NE LASER

6.4.1   Principle

A plate with semi-reflecting faces, is illuminated by a single-mode laser. Multiple-beam circular fringes of the plate are observed at infinity.

6.4.2   Setting-up the experiment (fig. 6.5)

a) Apparatus

S: He–Ne gas laser with a minimum output of 3 mW ($\lambda = 6328$ Å).

O: lens with a focal length of 8 cm.

L: semi-reflecting plate.

E: white screen.

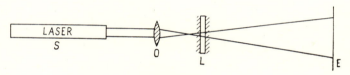

Figure 6.5

b) Adjustments

Distance between S and O: 10 cm.

Distance between O and L: arbitrary. The plate is normal to the light beam.

Distance between L and E: 3 to 4 meters, which is practically infinite.

6.5   FRINGES OF EQUAL THICKNESS

6.5.1   Principle

An air wedge of angle ε (fig. 6.6) formed between two plates, is illuminated by a parallel beam under normal incidence. The opposite faces of the two plates are plane and are highly reflecting.

The rays such as (1), (2), (3), ..., (p), originating from the same incident ray *SI*, interfere and give rise to interference fringes localised in the air wedge.

If the angle ε is so small that ε^2 could be neglected, the interference fringes represent the lines of equal thickness of the air wedge and are parallel to its edge, the fringe spacing being $\lambda/2\varepsilon$. If ε^2 is not negligible, the phenomenon is modified and the fringes no longer represent lines of equal thickness in the air wedge.

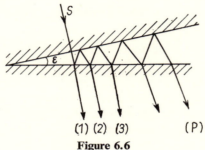

(1) (2) (3)   (P)

Figure 6.6

6.5.2 Setting up the experiment (fig. 6.7)

a) Apparatus

   S: mercury vapour lamp (typ A_3).

   O': lens with a focal length of 10 cm.

   L: two semi-aluminized plates held against each other in a wooden frame. An air wedge is formed by securing the two plates on one side by means of a rubber band.

   O'': lens with a focal length of 15 cm.

   O: lens with a focal length of 25 cm.

   E: white screen.

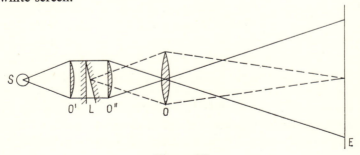

Figure 6.7

b) Adjustments

Distance between S and O': 10 cm. The light beam issuing from O' is parallel.

Distance between O' and L': about 15 cm. The plate L is so oriented that the beam falls on it normally.

Distance between L and O'': about 5 cm.

Distance between O'' and O: 15 cm. The lens O is placed at the image focus of O'', that is, at S', the image of S given by the composite system $O'O''$. The system L is imaged by O on the screen E, 4 metres away.

6.6   FRINGES OF SUPERPOSITION

6.6.1   Principle

Consider two plates with plane parallel, semi-reflecting faces, of thicknesses e and e', of refractive index n and making between them an angle ε (fig. 6.8). The thicknesses of the two plates differ very slightly. Two rays such as R and R' originating from the same incident ray interfere at infinity. The

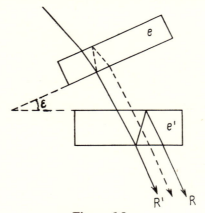

Figure 6.8.

interference fringes are equidistant straight lines, parallel to the edge of the wedge formed by the plates. These are the fringes of superposition of the two plates. The fringe spacing is $n\lambda/2\varepsilon e$ if the angle of incidence is small, ε is small and $e = e'$. It varies with ε: as ε decreases, the fringes broaden, and for $\varepsilon = 0$ uniform intensity is obtained over the whole field. The fringes are visible in white light. The central fringe is white with coloured fringes on either side. For $\varepsilon = 0$, we obtain a uniform colour which is white if $e = e'$.

6.6.2 Setting up the experiment (fig. 6.9)

a) Apparatus

S: carbon arc.

C: ordinary condenser.

L and L': two semi-reflecting plates obtained by evaporating 15 dielectric films of suitable refractive indices on two glass plates.

E: white screen.

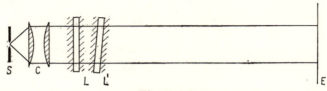

Figure 6.9

b) Adjustments

Distance between S and C: the source *S* is approximately at the object focus of *C*.

Distance between C and L: *L* is placed against the condenser.

Distance between L and L': *L* is held in hand to vary its inclination and is kept at about 5 cm from *L*.

Distance between L' and E: 4 to 5 meters which is practically equivalent to infinity.

7

Applications of Interferometry

7.1 MEASUREMENTS OF THERMAL EXPANSIONS

7.1.1 Principle

The interferometric method of Fizeau permits the measurement of the coefficient of linear expansion of a solid of small dimensions. A sample is cut in the form of a plane parallel plate normal to the direction along which the expansion is to be measured. The sample E (fig. 7.1) is placed on a plane surface P, and Newton's rings are formed by the light reflected from the upper face of E and from the lower face of a plano-convex lens L held by three adjustable spacers C. The whole assembly is placed in a thermostat. By varying the temperature, the variation in the thickness of E can be studied, taking due account of the variation of the index of refraction of air with temperature.

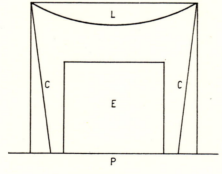

Figure 7.1

As a demonstration experiment, the expansion of a rod of dural may be shown with the help of a variant of the preceeding method. This rod is surrounded by a heating wire. When the current is passed through the wire, the rings broaden, the new ones appearing at the centre. On cooling the rod, the rings narrowed again, vanishing at the centre.

7.1.2 Setting up the experiment (fig. 7.2)

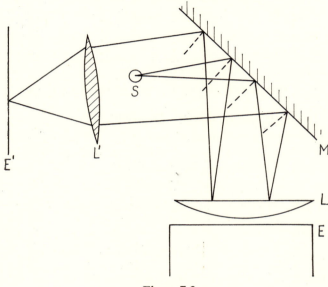

Figure 7.2

a) Apparatus

S: mercury vapour lamp (A_3) with a filter to isolate the green line.

M: plane mirror.

E: the sample in the form of a rod of 4 cm long and 3 cm in diameter. A standard piece of plane glass is pressed into contact with the polished surface of the sample. The upper face of the standard is lightly aluminized to increase its reflectivity.

L: plano-convex lens with its lower face lightly aluminized to increase the reflectivity. Its focal length is about one meter.

L': lens with a focal length of 18 cm.

R: heating wire (500 to 1000 Ω) worked by a rheostat under 70 to 80 volts.

E': white screen.

b) Adjustments

Adjustment of the mirror M: the mirror M receives the light beam coming from the source. It is inclined at 45°, sends the light vertically on the ensemble E and L and then receives the light reflected from this ensemble.

Adjustment of L' and E': the lens L' forms an image of E on the screen E'.

7.2   ANTI-REFLECTING THIN FILMS

7.2.1   Principle

Normally incident light illuminates a medium of refractive index n_3 separated from a medium of index n_1 by a film of thickness d and index n_2. The indices n_1, n_2 and n_3 are such that:

$$n_1 < n_2 < n_3 \qquad (7.1)$$

For $n_2 d = K\lambda/2$ (K being a whole number), the reflectivity of the medium of index n_3 is maximal and is equal to:

$$R_M = \frac{(n_1 - n_3)^2}{(n_1 + n_3)^2} \qquad (7.2)$$

For $n_2 d = (2K + 1)\,\lambda/4$ it is minimal and is equal to:

$$R_m = \frac{(n_1 n_3 - n_2^2)^2}{(n_1 n_3 + n_2^2)^2} \qquad (7.3)$$

If the medium n_3 is in direct contact with the medium n_1, its reflectivity in normal incidence is equal to R_M.

Thus by interposing between two media of indices n_3 and n_1 a thin film of index n_2 less than n_3 and of thickness d such that $n_2 d = (2K + 1)\,(\lambda/4)$, the reflectivity of the medium of index n_3 can be decreased. The reflectivity of a glass surface can be decreased by depositing on it a thin transparent film of magnesium fluoride ($n_2 = 1.35$) or of cryolite ($n_2 = 1.36$) of such thickness d that:

$$n_2 d = \lambda/4 \qquad (7.4.)$$

The experiment is conducted by receiving on a screen the light beam reflected from a glass plate only one half of which has been treated. We can thus observe that the reflectivity of the treated portion of the glass plate is less that of the non-treated portion.

7.2.2   Setting up the experiment (fig. 7.3)

a) Apparatus

S: carbon arc.

C: ordinary condenser (diameter 10 cm, focal length 5 to 7 cm).

M: plane mirror of about 15 cm in diameter.

L: glass plate half of which has been covered by slow evaporation in a vacuum with a film of magnesium fluoride ($n_2 = 1.35$) or of cryolite AlF$_3$, 3NaF ($n_2 = 1.36$). The back face has been blackened to avoid parasite light.

O: lens with a focal length of 30 cm.

E: white screen.

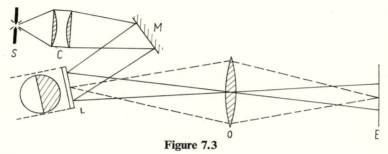

Figure 7.3

b) Adjustments

Distance between S and C: this distance is such that the light beam issuing from C should be very slightly converging.

Distance between C and M: about 35 cm.

Adjustment of M: the inclination of M with respect to the light beam coming from C is such that the light reflected from M falls normally on L.

Distance between M and L: about 35 cm.

Distance between L and O: lens O is placed in the plane of the image of S given by C after reflection by the mirror. L is close to the object focus of O and is imaged 4 or 5 meters away.

Distance between O and E: O forms an image of L on the screen E.

7.3   THIN FILMS TO INCREASE REFLECTION

7.3.1   Principle

If in the § 7.2.1 we suppose that $n_1 < n_3 < n_2$, we find that in normal incidence: R is maximal for $n_2 d = (2K + 1) \lambda/4$ and is given by:

$$R_M = \frac{(n_1 n_3 - n_2^2)^2}{(n_1 n_3 + n_2^2)^2} ; \tag{7.5}$$

R is minimal for $n_2 d = K\lambda/2$ and is given by:

$$R_m = \frac{(n_1 - n_3)^2}{(n_1 + n_3)^2} \qquad (7.6)$$

The reflectivity of a medium of index n_3 with respect to a medium of index n_1 can thus be increased by interposing between the two a film of index n_2 (n_2 is greater than n_3) and of thickness d such that:

$$n_2 d = (2K+1)\,\lambda/4 \qquad (7.7)$$

7.3.2 Setting up the experiment

The set up is the same as that of figure 7.3. The plate L is replaced by a plate L' one half of which has been treated by depositing on it a film of stibnite ($n \cong 3$) or of zinc sulfide ($n_2 \cong 2.3$) by slow evaporation in a vacuum.

7.4 INTERFERENCE FILTERS BY TRANSMISSION

7.4.1 Principle

A parallel beam of white light falls on a Fabry–Perot interferometer of thickness a at an angle of incidence i. Under these conditions, Φ (see 6.1) depends only on the wavelength. The interferometer lets pass only those radiations which satisfy the relation:

$$\Phi/2\pi = \delta/\lambda = K \qquad (7.8)$$

(K being a whole number).

If the transmitted light is analysed with a spectroscope, a channelled spectrum is observed consisting of sharp bright lines on a dark background. By a suitable choice of R and e, one can arrange to obtain only· one bright line of wavelength λ and of width

$$\Delta\lambda = \lambda\,\frac{1 - R}{\pi K \sqrt{R}}. \qquad (7.9)$$

When the angle of incidence increases, the transmitted radiation gets displaced towards blue (λ decreases). This is the principle of the interference filters by transmission. In practice, the air film is replaced by a transparent material (cryolite, lithium or magnesium fluoride) and the two semi-reflecting plates by multilayers of suitable thicknesses and indices so as to increase the reflectivity and to decrease the absorption to a minimum.

7.4.2   Setting up the experiment (fig. 7.4)

a) Apparatus

  S: carbon arc.
  C: ordinary condenser.
  F: vertical slit.
  L: lens with a focal length of 35 cm.
  F.I.: interference filter by transmission.
  P: direct vision prism.
  E: white screen.

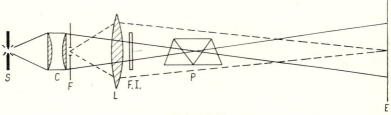

Figure 7.4

b) Adjustments

  Distance between S and C: this distance is such that the image *S'* of *S* is at about 40 cm from *C*.
  Distance between C and F: the slit *F* is placed against the condenser.
  Distance between F and L: the lens *L* gives an image of *F* on the screen *E*, about 4 m away.
  Distance between C and P: the direct-vision prism *P* is placed at *S'* so that all the light emitted by *F* passes through it.
  Position of F.I.: this is interposed between *L* and *P*.

7.5   INTERFERENCE FILTERS BY REFLECTION

7.5.1   Principle

The principle of interference filters by reflection is the same as that of the filters by transmission. Their theory is that of the Fabry–Perot interferometer used in reflection with a parallel beam of light. In practice, the air film is replaced by a thin layer of magnesium fluoride or cryolite, and the reflectivity of the back plate is optimised by metallisation.

7.5.2   Setting up the experiment (fig. 7.5)

a) *Apparatus*

S: carbon arc.
C: ordinary condenser.
F: vertical slit.
M: plane mirror.
F.I.: interference filter by reflection.
P: direct-vision prism.
L: lens with a focal length of 35 cm.
E: white screen.

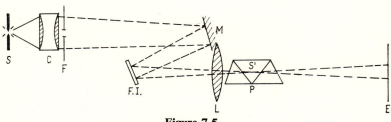

Figure 7.5

b) *Adjustments*

Distance between S and C: this distance is such that the light beam comnig out of *C* should be slightly convergent.

Distance between C and F: F is just close to *C.*

Distance between F and M: about 30 cm.

Distance between M and F.I.: about 30 cm. The light beam reflected from *M* must fall on *F.I.*

Position of P: P is placed at *S'* so as to receive whole of the light beam reflected by *F.I. S'* is the image of *S* as given by *C,* the light having been reflected by the mirror *M* and having traversed the lens *L.*

Position of L: the lens is as near to *S'* as possible and forms an image of *F* on the screen *E,* 4 metres away.

7.6   FILTERS FOR THE INFRA-RED REGION

7.6.1   Principle

The reflectivity and transmissivity of materials employed in the study of the infra-red region can be modified in the same way as outlined in 7.2

and 7.3 for the visible region. It suffices to deposit on the material to be used a thin film of suitable thickness and index of refraction: for example, the transmissivity of germanium can be increased from 41% to 90% by depositing on it a layer of SiO_2 of thickness $\lambda/4$. In the experiment to be performed, a photo-cell receives the light flux transmitted by an anticaloric filter (filter which lets pass the visible but stops the infra-red) and the light flux reflected by a cold mirror (filter which reflects the visible and lets pass the infra-red). The deflection of the galvanometer connected to the photo-cell is very weak in both the cases. It increases considerably if the cold mirror is replaced by an ordinary mirror or if the anti-caloric filter is removed.

7.6.2   Setting up the experiment with anti-caloric filter (fig. 7.6)

a) *Apparatus*

   S: carbon arc.
   C: ordinary condenser.
   F: filter absorbing the infra-red.
   R: photo-cell sensitive to infra-red radiation, connected to a galvanometer *G*.

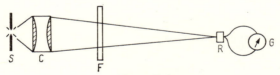

Figure 7.6

b) *Adjustments*

   Distance between S and C: the source *S* is very close to the condenser *C* so that the image *S'* is at about 50 cm from *C*.
   Distance between C and R: the photo cell is placed at *S'*.

7.6.3   Setting up the experiment with a cold mirror (fig. 7.7)

The set up is the same. The filter is replaced by a mirror *M* which is treated to obtain a minimum of reflectivity for the infra-red. The light emitted by the source *S* is converged by the condenser *C*, is reflected by the mirror *M* and comes to focus at *R*.

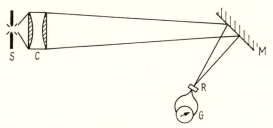

Figure 7.7

7.7 LIPPMANN PHOTOGRAPHY

7.7.1 Principle

A sensitive layer of extremely fine grains of silver chloride, deposited on a mirror, is illuminated in white light. Each radiation composing the incident light gives rise to a system of stationary waves. The silver chloride is reduced in the anti-nodal planes and thus parallel equidistant layers of silver are formed on the photographic plate, the distance between two consecutive layers being $\lambda/2$.

In short, at a point of the photographic plate, we obtain a system of reflecting layers characteristic of the radiations received at that point. The photographic plate is developed and illuminated by white light incident normally; the reflected light is observed. The light reflected by each point of the plate reproduces the colour which was used to expose it.

This method, abandoned for the fabrication of colour photographs, is now used in holography. Colour holograms are, in fact, obtained by an analogous technique.

7.7.2 Setting up the experiment (fig. 7.8)

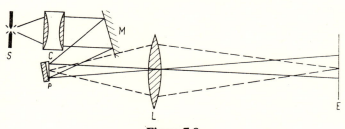

Figure 7.8

The arrangement is the same as that of figure 3.4 except that the element A is replaced by a Lippmann photographic plate. This plate is covered with a glass wedge (angle of a few degrees) to eliminate the reflected light from the back surface of the emulsion.

7.8   FILTERS BY REFLECTION FOUND IN NATURE

7.8.1   Principle

Certain crystals such as potassium chlorate, certain coleopter and certain birds (colibris) show bright colours when they are observed in reflected light. The colours change with the angle of observation. This is due to the fact that the carapace of these insects and the feathers of these birds consist of thin multi-layers of alternating high and low indices of refraction.

7.8.2   Setting up the experiment

A carbon arc illuminates (fig. 7.9) a naturalised colibri A or a beetle. Different colours are observed depending upon the angle of incidence and the angle of observation.

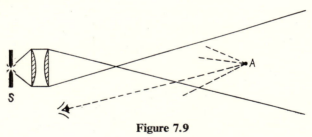

Figure 7.9

8

Diffraction at a finite Distance (Fresnel)

8.1 DIFFRACTION AT THE EDGE OF A SCREEN

8.1.1 Principle

The screen E, a semi infinite plane bounded by a straight edge, is set up normal to the light rays issuing from a point source S (fig. 8.1). In the plane P one observes the geometrical shadow of the screen E and a series of diffraction fringes parallel to the edge of the screen in the illuminated region.

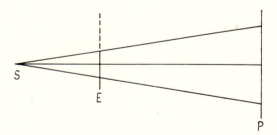

Figure 8.1

If the relative intensity I/I_0 at a point on P is represented as a function of the distance of this point from the edge of the geometrical shadow, we obtain the curve shown in figure 8.2.

On bringing the screen of observation P nearer to the screen E, a rapid decreases in the fringe width is observed. At a few centimeters from E, it is necessary to use a powerful eyepiece to see the fringes.

65

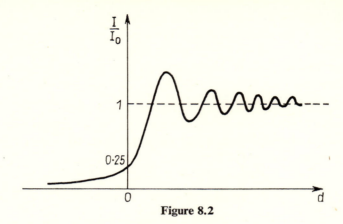

Figure 8.2

8.1.2   Setting up the experiment (fig. 8.3)

a) Apparatus

   S: high pressure mercury vapour lamp A_3 or A_6, provided with a filter to isolate the green line.

   C: ordinary condenser.

   F: vertical slit of adjustable dimensions.

   E: screen with a good straight edge made of blackened thin cardboard or of a metal plate.

   R: television camera, vidicon, without the objective, related to the television receiver on which the diffracted image appears.

   Filters: filters are used to attenuate the light falling on the photo-cathode of the camera so that it is not saturated.

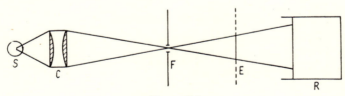

Figure 8.3

b) Adjustments

   Distance between S and C: this distance is such that the condenser forms an image *S'* of *S* at about 5 cm.

   Distance between C and F: the vertical slit is placed at *S'*.

Distance between F and E: about 40 cm. A tube is placed between the slit and the edge of the screen so as to prevent parasite light from reaching the photo-cathode. The slit and the straight edge must be rigorously parallel.

Distance between E and R: about 30 cm. A tube is placed between *E* and *R* also to avoid parasite light.

c) Note

The distances indicated are approximate. The straight edge and the photo-cathode should be so adjusted that the diffraction pattern on the screen of the television is as bright as possible.

8.2   DIFFRACTION BY A THREAD

8.2.1   Principle

The diffraction pattern produced by a thread (fig. 8.4) is studied by the same arrangement as that of fig. 8.1. A series of bright and dark fringes is observed on the screen *P*.

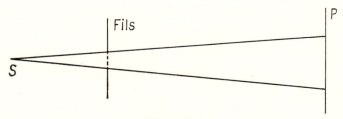

Figure 8.4

8.2.2   Setting up the experiment (fig. 8.5)

The set up is the same as that of figure 8.3. *E* is replaced by a series of threads of different diameters. The parallelism between the slit *F* and the threads must be accurately adjusted.

Figure 8.5

8.3 DIFFRACTION AT A CIRCULAR APERTURE

8.3.1 Principle

Consider a point source S, a circular hole in an opaque screen D placed in front of the source and an observation screen P (fig. 8.6). The straight line joining the source S to the centre O of the hole is normal to the opaque screen D. The diffraction pattern observed at P has the symmetry of revolution about the line SOO'. A series of more or less bright and dark rings is observed. The centre of the pattern is bright or dark depending on the geometry of the experiment.

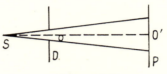

Figure 8.6

If the diffracting hole is replaced by a circular opaque screen, bright and dark rings are again seen on the screen P; however, the centre of the diffraction pattern is always bright.

8.3.2 Setting up the experiment

The experimental arrangement is the same as that of figure 8.3 except that the slit F is replaced by a circular aperture of as small a diameter as possible (a point source) and the straight edge E is replaced by a circular hole or an opaque circular screen.

An opaque circular screen can be made by fixing a ball-bearing to a thin glass plate by means of gum or sealing wax.

8.4 HOLOGRAPHY

8.4.1 Principle

Interference is produced between the light diffracted by an object and a coherent background (fig. 8.7). The coherent background light makes an angle θ with the mean direction of diffracted light. The resultant intensity is recorded by a photographic plate H. After developing, the plate is illuminated by a parallel beam incident normally. Three beams emerge from H (fig. 8.8)

1) a non-deviated beam (1).

2) a beam (2) deviated in the direction $+\theta$ which gives a real image O' of the object.

3) a beam (3) deviated in the direction $-\theta$ which gives a virtual image O'' of the object.

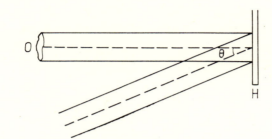

Figure 8.7

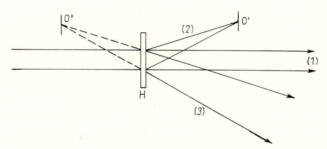

Figure 8.8

When the image O'' is observed, through the hologram H, it is noticed that the image reconstructs the object in three dimensions. The reconstruction in three dimensions obtained with a hologram must not be confused with subjective relief. The experiment consists of televising the virtual image O'' and thus showing that it reconstructs the three dimensions of the object.

By observing the television screen, it is noticed that the eye cannot be focused simultaneously on different parts of the object at the same time. If the camera is rotated about a mean position of observation, the object is seen from different angles. The regions masked formerly, may appear in view. Clearly this phenomenon cannot be obtained in an ordinary photograph.

8.4.2   Setting up the experiment (fig. 8.9)

a) Apparatus

S: He-Ne gas laser, output 3 mW.

O: microscope objective of magnification × 10.

H: hologram.

R: television camera, Vidicon, with objective.

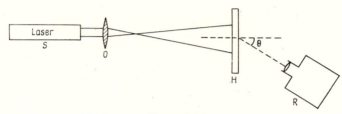

Figure 8.9

b) Adjustments

Distance between S and O: arbitrary.

Distance between O and H: find it by trial since it depends upon the conditions under which the hologram was constructed. The objective O is used to enlarge the laser beam so that it illuminates the entire hologram.

Orientation of H: easily obtained by trial. It depends also upon the conditions under which the hologram was constructed.

Variation

If a laser source and the television equipment are not available, the hologram H can be observed (fig. 8.9) individually by using a mercury vapour lamp, type A_3, as a source. With a 1 mm hole placed in the focal plane of a lens, 30 to 40 cm of focal length, a quasi-parallel beam is obtained covering the entire surface of H. The eye is placed in the position of R at about 30 cm from H. By isolating with the help of suitable filters, the green or yellow line of mercury, it can be shown that the magnification depends on the wavelength of light employed for observation.

9

Diffraction at Infinity (Fraunhofer)

9.1 DIFFRACTION AT A NARROW SLIT

9.1.1 Principle

A slit source S, placed in the focal plane of a lens L_1, gives a parallel beam of light (fig. 9.1). This light beam falls on a screen having a narrow slit F parallel to the source S. A lens L_2 is placed against the slit F. The diffraction pattern of the slit F is observed in the focal plane P o fL_2. It consists of a series

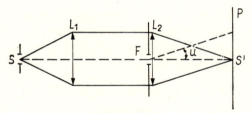

Figure 9.1

of diffraction fringes alternatively bright and dark, parallel to the slit source and symmetrical with respect to SS'. The central bright fringe is located at S', the geometrical image of S. The distribution of intensity in the plane P is given by the formula:

$$I = I_0 \left(\frac{\sin kua}{kua} \right)^2 \tag{9.1}$$

where $K = 2\pi/\lambda$, $2a$ is the width of the slit F and u is the sine of the angle of diffraction in the plane of the figure 9.1 (fig. 9.2).

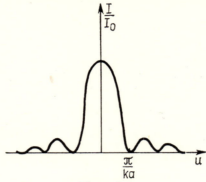

Figure 9.2

a) the experiment shows that the fringes get closer to the central fringe as the slit F is widened. If the slit F is made finer, the diffraction pattern spreads out.

b) By varying the height of the slit source S, the height of the fringes varies in the same sense.

c) By rotating the slit source in its plane, the fringes are rotated in the same sense and remain equally sharp. Only a fine bright line is observed when the slit source is perpendicular to the slit F.

d) If the slit F is displaced in its own plane keeping it parallel to its original direction, the diffraction pattern remains stationary. The phases of all the vibrations arriving at a given point in the plane P change by the same amount but the intensities undergo no change.

9.1.2   Setting up the experiment (fig. 9.3)

a) Apparatus

$S:$ mercury vapour lamp of high pressure, type A_3, with a filter isolating the green line.

$C:$ ordinary condenser.

F_1 *and* $F_2:$ two vertical slits of adjustable heights and widths.

$O:$ lens with a focal length of about 30 cm and of any diameter since a very small aperture will be used.

$R:$ television camera (Vidicon) without the objective, related to the television receiver on the screen of which the diffracted image will appear.

Filters: neutral filters are used to attenuate the light beam so that the photo-cathode of the camera does not get saturated with light intensity at the centre of the diffraction pattern.

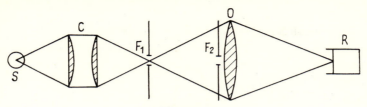

Figure 9.3

b) Adjustments

Distance between S and C: arbitrary, provided that a real image of *S* is formed at F_1.

Distance between C and F_1: the slit F_1 is so placed that the image of *S* falls on it.

Adjustment of O and of R: the positions of the objective and of the photocathode of the camera are adjusted by trial and error so as to obtain a good image of the slit source on the television screen. This image must be uniform and bright. During this adjustment, the slit should be almost closed so as not to saturate the photo-cathode.

Adjustment of the slit F_2: the diffracting slit is placed against the objective. The width of the slit source must be increased till the phenomenon is properly visible. The width of the slit F_2 is so adjusted that the diffraction fringes on either side of the central image become visible.

9.1.3   Variation (fig. 9.1)

a) Apparatus

S: He-Ne laser ($\lambda = 6328$ Å) with a power of about 15 mW.

O: microscope objective $\times 10$

D: a hole of about 25 microns diameter.

F: vertical slit of adjustable width and height.

E: white screen.

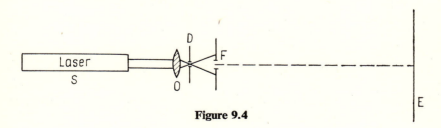

Figure 9.4

b) Adjustments

Distance between S and O: arbitrary.

Distance between O and D: the circular hole is at the image focus of *O*.

Distance between D and F: arbitrary.

Distance between F and E: 4 to 5 m.

c) Remark

If the slit *F* is illuminated directly by the laser beam, a very bright line is observed on the screen; this represents the diffraction with a point source.

9.2 DIFFRACTION AT A CIRCULAR APERTURE

9.2.1 Principle

A point source *S* (fig. 9.5) placed at the object focus of a lens L_1 gives a parallel beam of rays. This beam falls on an opaque screen E_1 having a circular aperture *T*. The diffraction pattern of the circular aperture (Airy's disc) is observed in the focal plane of the lens L_2.

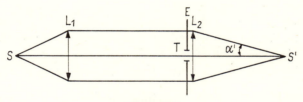

Figure 9.5

At *S'* one observes a very bright disc encircled by dark and bright rings (fig. 9.6). The distribution of intensity in the plane *P* is given by the formula:

$$I = I_0 \left[\frac{2J_1(Z)}{Z} \right]^2 \tag{9.2}$$

I_0 = intensity at the centre of the diffraction pattern.

$Z = (2\pi/\lambda)\varrho\alpha'$

α' = angular aperture of L_2.

ϱ = radius of L_2.

J_1 = Bessel's function of first order.

a) By varying the diameter of the hole *T*, it is shown that the diameter of the diffraction pattern varies inversely as the diameter of the hole.

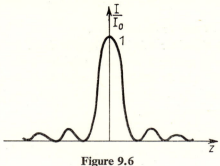

Figure 9.6

b) On displacing the aperture T in its own plane, it is shown that the diffraction pattern changes neither in intensity nor in position (only the phases change).

c) If the screen P is displaced in the direction SS', the diffraction patterns with defects of focus are observed. In the case when the lens has aberrations, these modify the structure of the diffraction pattern.

9.2.2 Observation of Airy's disc (fig. 9.7)

a) Apparatus

S: He-Ne laser with a power of about 0.3 mW.

O: lens of 35 cm focal length and of arbitrary aperture since it will be used with a very small aperture.

R: television camera (Vidicon), without the objective, connected to television receivers on which the diffraction pattern appears.

T: hole made in a tin foil of about 0.1 mm thickness. Make different trials with a needle till the diffraction pattern (6 to 7 rings) covers approximately half of the television screen.

Filters: use the neutral filters so that the photocathode of the television camera does not get saturated by the intense central disc. The adjustment has to be done by experiment. The filters will be placed at the exit of the laser.

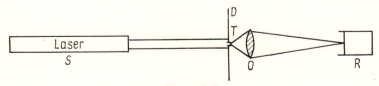

Figure 9.7

b) Adjustments

Distance between S and T: 60 to 70 cm.

Distance between T and O: *T* is in contact with *O*.

Distance between O and R: The photocathode of the camera lies in the focal plane of *O*. A tube of black paper should preferably be placed between the lens and the photocathode so as to prevent the diffused light from reaching the photocathode.

e) Comment

With a sufficiently powerful laser (at least 20 mW) Airy's pattern can be directly projected on a screen. The lens *O* is suppressed and a hole *T* of 2 to 3 mm in diameter is used. The phenomenon may be observed at a distance of 5 to 6 metres.

9.2.3   Observation of Airy's pattern in the presence of defect of focus and aberrations (fig. 9.8)

a) Apparatus

S: He-Ne laser with a power of about 0.3 mW.

M: Microscope consisting of an objective × 100, an eyepiece × 8 or more and a condenser *C* (ordinary Abbe's condenser).

A: object slide with its upper surface (facing the objective of the microscope) aluminized by evaporation. One always finds in the aluminium small irregularities having diameters less than that of the diffraction disc

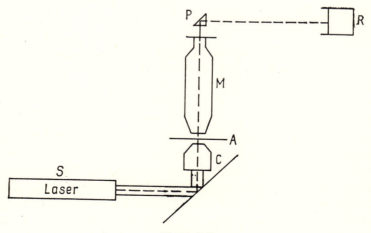

Figure 9.8

of the microscope objective. These constitute point sources enabling the formation of beautiful diffraction patterns.

A drop of immersion oil is interposed between the aluminized object slide and the objective × 100. To see the influence of the spherical aberration on the diffraction figure, it suffices to omit the oil.

R: Television camera (Vidicon) without the objective. In order to be able to use the camera in the horizontal position, a total reflection prism *P* or a mirror is placed immediately after the eyepiece of the microscope to send the beam in the horizontal direction.

b) Adjustments

Distance between S and M: arbitrary.

Adjustment of the microscope: focus the microscope on the object and search for the diffraction pattern with a characteristic aspect. Adjust the distance of the condenser from the object such that the diffraction figure is as bright as possible.

Distance between the microscope eyepiece and the photocathode: about 10 cm. A tube of black paper should preferably be placed between the eyepiece and the photocathode to prevent parasite light from reaching the photocathode.

Use of a source other than a laser

The laser may be replaced by a high pressure mercury vapour lamp infront of which a green filter is placed. A lamp of A_3 type is quite suitable for this experiment.

9.3 DIFFRACTION BY A LARGE NUMBER OF SMALL OPAQUE SCREENS

9.3.1 Principle

The figure illustrating the principle is similar to the preceeding figures (fig. 9.9). A screen *E* having a large number *N* of identical apertures, oriented in the same way and distributed irregularly, is placed before a lens L_2. These apertures may be circular, for example. In the focal plane *S'* of the lens L_2, one obtains diffraction figure identical to that obtained with a single aperture but *N* times more luminous. If the *N* apertures are replaced by *N* opaque screens whose contours are identical to those of the apertures, one obtains the same figure of diffraction at *S'* (theorem of Babinet).

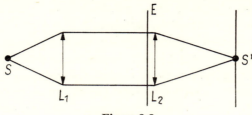

Figure 9.9

9.3.2   Setting up the experiment (fig. 9.10)

a) Apparatus

S: carbon arc.

C: ordinary condenser.

D: diaphragm with circular aperture of different dimensions.

O: corrected lens with a focal length of 25 cm.

E: plate with small opaque discs. This may be obtained as follows. A plane parallel of glass is lightly oiled on one face and is then rubbed with a piece of cloth. Lycopodium powder is then blown over it so that the particles of powder are deposit uniformly on the glass plate. The plate is then shaken to remove loose powder.

R: white screen.

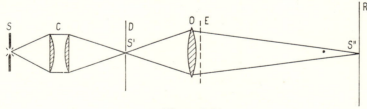

Figure 9.10

b) Adjustments

Distance between S and C: this distance is such that the condenser forms an image of S at a distance of 40 cm.

Distance between C and D: D is placed at *S'*. The diameter of the aperture is adjusted to obtain sharp rings. If the hole is too large, the phenomenon becomes integrated and the sharpness diminishes.

Distance between D and O: D is very near to the object focus of *O*, so that the image *S''* of *S'* is formed on the screen *R*, far away.

Distance between O an E: a few centimetres.

Distance between E and R: the observation screen is at S'', the image of S' given by O, at a distance of about 4 metres from O.

9.4 DIFFRACTION BY A FINITE NUMBER OF EQUALLY SPACED SLITS

9.4.1 Principle

In the scheme of figure 9.1, the diffracting slit F is replaced by n identical and equally spaced slits parallel to S. The intensity in the focal plane of L_2 is given by the formula:

$$I = I_0 \left[\frac{\sin kua}{kua} \right]^2 \left[\frac{\sin nkub}{nkub} \right]^2 \qquad (9.3)$$

where $2a$ is the width of each of the slits, $2b$ is the distance between the centre of two consecutive slits, u is the direction cosine of the vibration diffracted in a plane normal to the direction of slits, $K = 2\pi/\lambda$.

The diffraction pattern represented by the first term of the product which is the same for each slit, is modulated by the interference phenomenon between the vibrations diffracted in the same direction by each of the slits. The interference phenomenon is represented by the second term of the product. Thus one observes interference fringes in the interior of the bright diffraction fringes.

a) $n = 2$. This is the arrangement for observing Young's fringes (fig. 9.11).

The bright fringes are called principal maxima.

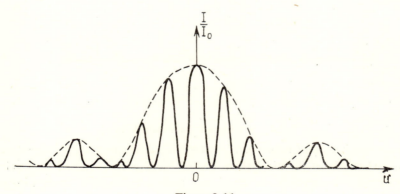

Figure 9.11

b) $n = 3$. Between two principal interference maxima, there appear the secondary maxima (fig. 9.12). The positions of principal maxima remain unchanged.

c) As n is increased, the principal maxima become sharper without getting displaced and the secondary maxima become larger in number and get weaker (fig. 9.13).

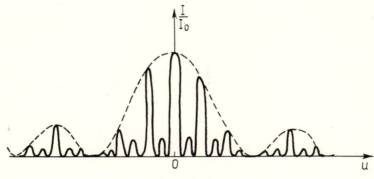

Figure 9.12

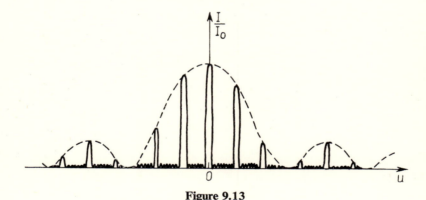

Figure 9.13

9.4.2   Setting up the experiment

The scheme is that of figure 9.3, in which the diffracting slit F_2 is replaced by an ensemble of identical slits. These slits can be obtained by tracing lines on an over-exposed and developed photographic plate or by winding a thread on two screws.

9.5 GRATINGS

9.5.1 Principle

A plane line grating is a transparent or reflecting plane surface on which parallel and equidistant grooves have been traced. It is along these grooves that the transmissivity or the reflectivity has been modified. The distance between two consecutive grooves is called the period of the grating. The number of grooves is always of the order of a few thousand.

When a parallel beam of light falls on the surface of a grating, a part of the incident light is transmitted without deviation (or regularly reflected); however, the other part finds itself in well determined directions along which the light vibrations diffracted by each of the grooves are in phase and where constructive interference is produced. Let F_1 and F_2 (fig. 9.14) be two consecutive grooves of the grating; a is the distance between them, i is the angle

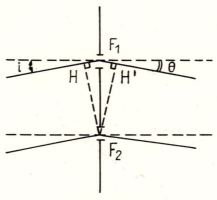

Figure 9.14

of incidence and θ is the direction in which the diffracted light is observed. The path difference between vibrations diffracted by F_1 and F_2 is;

$$\delta = F_1 H + F_1 H' = a(\sin i + \sin \theta) \qquad (9.4)$$

where the same sign has been given to i and θ since the direction of incidence and the direction of observation are on the same side of the normal to the grating. The relation (9.4) is valid, with the same conventions, for a reflection grating. The condition $\delta = \pm K\lambda$ (K being an integral number) defines the directions θ_m of constructive interference. The value of K gives the order of interference. $K = 0$ corresponds to $\theta = -i$, that is, to the beam transmitted without deviation (or regularly reflected). If the incidence is normal, we have:

$$\sin \theta_m = \pm \, K(\lambda/a) \tag{9.5}$$

When the source emits polychromatic light, each component monochromatic line gives maxima in the directions given by (9.5) and these are different except for $K = 0$. One, thus, obtains a spectrum of the source. The spectral lines of longer wavelengths correspond to larger values of θ_m.

9.5.2   Setting up the experiment (fig. 9.15)

a) *Apparatus*

S: carbon arc.
C: ordinary condenser.
F: slit with adjustable width and height.
L: lens of about 200 cm focal length and of large diameter.
R: plane reflection grating having about 400 lines per millimeter.
E: white screen.

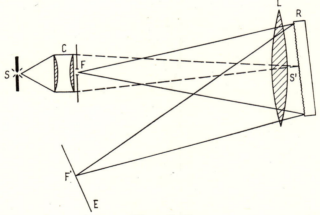

Figure 9.15

b) *Adjustments*

Distance between S and C: this distance is such that the image S' of S given by the condenser should be about 1 m away from C.

Distance between C and F: the slit F is placed against the condenser. It is parallel to the lines of the grating.

Distance between F and L: about 90 cm.

Distance between L and R: the grating R is placed at S'.

Distance between R and E: the light reflected by the grating passes again through the lens L. This lens forms an image F' of F at a distance of about 4 m. The screen is placed at F'.

9.5.3 Usage of a grating

The relation (9.4) can be satisfied for $K = 1$ only when $\lambda < 2a$. Thus for a grating of a given period, there exists a maximal wavelength above which the light will not appear in the first order spectrum. But a grating with a period very much greater than λ, can be used under grazing incidence. It is in this way that a grating made for the visible region can be employed for diffracting X-rays. The following experiment gives an analogy of this process. A slit placed against a mercury vapour lamp illuminates a metallic scale graduated in millimetres. When the incidence is nearly grazing, the spectrum of the source is observed in the reflected light.

9.6 A CROSS-RULED SCREEN IN WHITE LIGHT

9.6.1 Principle

A cross-ruled screen is a two dimensional grating. In the set up of figure 9.15, the slit source is replaced by a circular hole and the line grating by a two dimensional grating of square mesh of side b. The diffraction pattern consists of an ensemble of images the centers of which form a grating of square mesh with side proportional to $1/b$. The relative intensities of various images are represented by the diameter of the circles on figure 9.16. In white light, each image is replaced by a spectrum.

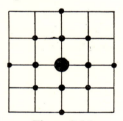

Figure 9.16

9.6.2 Setting up the experiment (fig. 9.17)

a) Apparatus

  S: carbon arc.

  C: ordinary condenser.

  D: diaphragm with holes of different diameters.

T: ordinary crossed grating or fine muslin or silk streched in a frame so that the threads remain straight. One may use two line gratings with period of 20 μ or 30 μ and putting them one against the other with the lines crossed.

L: lens with a focal length of 25 cm.

E: white screen.

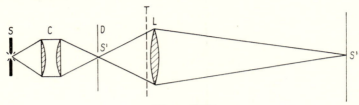

Figure 9.17

b) Adjustments

Distance between S and C: arbitrary, provided that *C* forms a real image *S'* of *S* on the hole of the diaphragm *D*. The light beam must cover the lens *L*.

Distance between D and L: slightly over 25 cm so as to obtain an image *S''* of *S'* at a distance of about 4 m on *E*.

Position of T: close to *L*.

9.7   PHASE CONTRAST AND THE DARK GROUND METHOD

9.7.1   Principle

In figure 9.18 the point source *S*, placed in the focal plane of a lens illuminates a transparent object *P*, presenting small variations of optical thickness extending over a small area. The image *P'* of *P* is observed in the conjugate plane with respect to the lens 0.

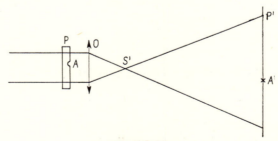

Figure 9.18

The structure of the plane wave-front is modified, on traversing the object, by the irregularities of the optical path. Let φ be the change in phase characterizing a defect of optical thickness at the point A. If φ is small such that φ^2 can be neglected, the wavefront which has traversed the object can be broken down into two waves.

— a vibration of amplitude a, not affected by the irregularity A of the object.

— a vibration of amplitude $a\varphi$ in phase quadrature with the first vibration and characterizing the defect of the object at A.

To these two vibrations correspond two beams:

1) direct beam not affected by the irregularities of the object P. It has the amplitude a and is represented by (1) in figure 9.19. It converges to S', the image of S.

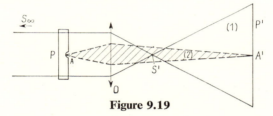

Figure 9.19

2) The beam diffracted by an irregularity such as A. It has the amplitude $a\varphi$ and is represented by (2) in figure 9.19 (hatched beam). It converges to A' the image of A. The vibrations of the diffracted beam are out of phase by $\pi/2$ with respect to the vibrations of the direct beam.

a) The phase contrast method consists in rendering the vibration of the direct beam in phase with the vibration of the diffracted beam. This is done by placing at S' a transparent plate called "the phase plate". The plate should be very small so that only the direct beam passes through it. In fact, it may be assumed that the beam diffracted by the small irregularities is very much spread out at S' (see fig. 9.2). Thus it passes almost entirely outside the phase plate. The phase plate must have such an optical thickness that it introduces a phase variation of $\pi/2$ in the beam traversing it with respect to the beam that passes outside. In this way we obtain at A' two vibrations which are in phase and have amplitudes a and $a\varphi$. They interfere and the intensity is proportional to:

$$a^2(1 + \varphi)^2 \cong a^2(1 + 2\varphi). \tag{9.6}$$

The contrast is $\gamma = 2\varphi$. Thus the phase variations of the object have been transformed into variations of intensity in the image P'. The transparent object becomes visible. If φ is small, so is also the contrast 2φ. The contrast can be considerably increased by rendering the phase plate absorbing.

b) The dark ground method consists in placing a small opaque screen at S' to stop the direct beam. At P' only the diffracted beam of amplitude $a\varphi$ is observed. The intensity is proportional to the square of φ and the phase variations appear bright on a dark ground. This method does not reveal the sign of φ and is less sensitive.

9.7.2   Setting up the experiment (fig. 9.20)

a) *Apparatus*

S: high pressure mercury vapour lamp (A_3 type) with a filter isolating the green line.

C: ordinary condenser.

F: vertical slit of adjustable width.

O_1: lens of 1 m focal length, the polishing defects of which are to be observed by phase contrast.

L: the phase plate consisting of a plane parallel glass plate, half of which is covered with an aluminium film of density equal to two. The dimensions of the phase plate are approximately 3 cm × 5 cm (fig. 9.21). With an aluminium film of this density, the required phase difference of $\pi/2$ is automatically obtained.

O_2: a lens of 35 cm focal length.

E: opaque screen having the same dimensions as the phase plate (a piece of card board for example). This screen replaces the phase plate when observation is to be made with the dark background method.

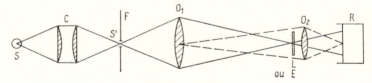

Figure 9.20

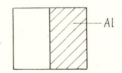

Figure 9.21

R: television camera (Vidicon) without the objective, connected to television receivers on which the phenomenon is observed.

b) Adjustments

Distance between S and C: this distance is such that *C* forms an image *S'* of *S* on *F*.

Setting up of O₁ and L: the lens O_1 forms an image of *F* on *L*; this image should be at about 3 m from *F*. *L* is so placed that the image of *F* is parallel to the line of separation of glass and aluminium, the image of *F* being on the aluminium. Under these conditions, half of the diffracted light is lost but the structure of the image is not modified.

Setting up of O₂ and R: O_2 is placed against *L* and forms an image of O_1 on the photocathode of the camera. This image should have a diameter of about 10 mm.

9.7.3 Another experiment using the dark ground method (fig. 9.22)

a) Apparatus

S: carbon arc.

C: ordinary condenser.

A: glass cell filled with water to which a few drops of glycerine have been dropped.

L: lens of 30 cm focal length.

D: opaque screen (a piece of cardboard for example) of such dimensions as to cover the image of *S'* given by the condenser.

E: white screen.

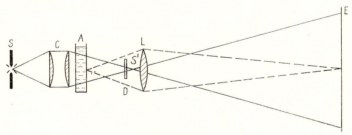

Figure 9.22

b) Adjustments

Setting up of S, L and C: the condenser forms an image *S'* of *S* on *L*.

Setting up of A, L and E: *A* is placed close to *C* and its image is formed by *L* on *E* placed 3 to 4 m away.

7*

9.8 FILTERING OF SPATIAL FREQUENCIES IN COHERENT ILLUMINATION: ABBE'S EXPERIMENT

9.8.1 Principle

In Abbe's experiment, a grating R is illuminated by a point source at infinity (fig. 9.23). The lines of the grating act as coherent sources and emit vibrations which interfere. In the focal plane of a lens L_2, placed after the grating, one observes the central image S'_0 (fig. 9.24) and the spectra of different orders, S'_1, S'_{-1}, ...

A lens L_3, placed in the plane of the spectra, forms an image of the grating in the plane P. It is shown that by suppressing certain diffraction images in the plane of L_3, the image R' observed at P can be considerably modified.

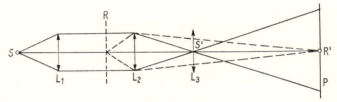

Figure 9.23

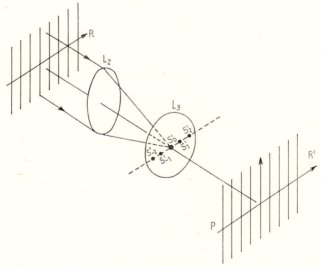

Figure 9.24

For example, if by means of a suitable diaphragm, the odd order images are suppressed, the image R' will have twice the number of lines as the actual grating R. If all the images except for the central one S'_0 are suppressed, the image R' is uniformly illuminated and it appears as if the grating R were not present.

9.8.2   Setting up the experiment (fig. 9.25)

a) *Apparatus*

S: He-Ne laser of about 15 mW power.
O: microscope objective.
T: cross grating of period $^1/_{10}$ mm.
D: a screen of suitable form to mask the spectra of the source.
E: white screen.

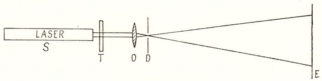

Figure 9.25

b) *Adjustments*

Distance between S and O: arbitrary.
Distance between S and T: T is placed between *S* and *O*.
Position of D: D is in the image focal plane of *O*.
Position of E: E is placed at the geometrical image of *T* as given by *O*.

Variation

If a laser is not available, the experiment can be conducted in the following way (fig. 9.26); the periodic object R is a two dimensional grating (metallic sieve). The slit F (parallel to one set of lines of the grating and in the plane of which the diffraction pattern of the source is formed), is narrowed till the spectrum of zero order alone is transmitted; under these conditions, the image of the lines parallel to the slit F is suppressed.

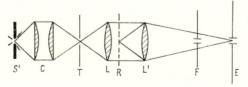

Figure 9.26

a) Apparatus

　S: carbon arc.

　C: condenser.

　T: circular aperture of 2 to 3 mm diameter.

　R: metallic sieve, 5 cm × 5 cm, of square mesh with sides of the order of 1 mm.

　L and L': lenses, focal length is about 30 cm.

　F: adjustable slit.

　E: white screen

b) Adjustments

　Distance between S and C: C forms an image of *S* at *T.*

　Distance between L and T: T is at the object focus of *L.*

　Distance between L and R: arbitrary.

　Distance between R and L': L' forms a real image of *R* at *E.*

　Distance between L' and F: F is at the object focus of *L'.*

　Distance between F and E: a few meters.

9.9 DIFFRACTION OF LIGHT BY ULTRASONIC WAVES

9.9.1 Principle

The elastic longitudinal, monochromatic and quasi-plane waves propagating in a liquid, create variations of density and consequently variations of refractive index. These variations are periodic in space and time. The medium is thus transformed into a phase grating with which Fraunhofer diffraction phenomena can be produced, provided that the period of the grating, equal to the wavelength of the elastic waves, is of the order of the wavelength of light. The ultrasonic waves are produced by a piezo-electric quartz plate of known frequency. When their amplitude is small, the variations of the refractive index are sinusoidal and the spectra of order 1 alone are observed. When these periodic variations are no longer sinusoidal, the different terms of the harmonic analysis give rise to spectra of higher orders.

9.9.2 Setting up the experiment (fig. 9.27)

a) Apparatus

　S: carbon arc.

　C: ordinary condenser.

　F: slit.

L: projecton lens, $f = 20$ to 25 cm.

K: cell with parallel faces filled with xylene (about 10 cm × 5 cm × 3 cm).

Q: piezo-electric quartz of frequency 4 Megahertz, vibrating on third harmonic and working on an oscillator to be described.

E: white screen.

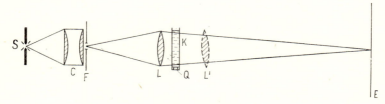

Figure 9.27

b) Optical adjustments

Distance between S and C: arbitrary.

Distance between C and F: F is against *C.*

Adjustment of F, L and E: L forms an image of *F* on the screen *E* situated 3 to 5 meters away.

Position of K: K is against *L.*

c) Electrical adjustments

The figures 9.28 and 9.29 give the circuit diagrams of the oscillator. The frequency can be varied between 8.3 and 14 MHz.

d) Comment

The phase grating can be rendered visible on the screen *E* by the dark ground method. The lens *L* gives a parallel beam of light. A lens *L'* ($f = 30$ to 40 cm) is placed after the cell *K* and forms an image of *K* on *E.* The image of *F,* formed in the focal plane of *L',* is masked by a screen. A system of lines alternatively dark and bright is observed at *E.* The period of the grating is equal to half the wavelength of elastic waves.

9.10 DIFFUSION OF LIGHT

9.10.1 Principle

A parallel beam of monochromatic light, of wavelength λ, propagates along the direction Ox (fig. 9.30) in a material medium containing isotropic dielectric particles of volume V. The electric field of the light wave induces

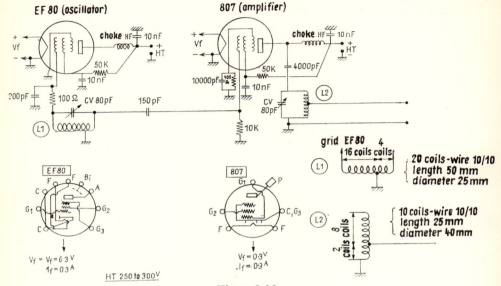

Figure 9.28

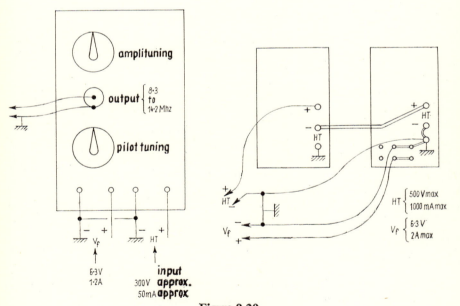

Figure 9.29

in each particle an electric dipole moment M of the same frequency. If the particles are isotropic, transparent and small in comparison to λ, we have:

$$M = \chi VE.$$

x is the electric susceptibility of particles and is equal to the induced moment in a unit volume by a unit field. The induced dipoles radiate in all directions, as the oscillators of Hertz. In the direction Ox, the composition of wavelets radiated with the incident wave takes into account the refraction of the medium. In a lateral direction, there is the diffused light. The calculation gives the following result: let $\mathscr{E}$ be the light intensity produced on a plane

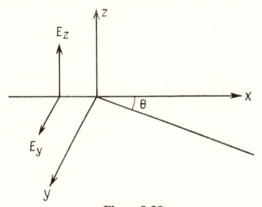

Figure 9.30

surface normal to Ox by the incident wave supposed to be formed of natural light. In a direction making an angle θ with Ox and lying in plane xOy, one observes the superposition of the effects of the two components E_y and E_z of the incident field, incoherent with each other. The intensity diffused by a unit volume of the medium of index n, containing N particles, is given by:

$$I = \frac{9\pi^2 N V^2}{2\lambda^4} \left(\frac{n^2 - 1}{n^2 + 2} \right)^2 (1 + \cos^2\theta)\mathscr{E}.$$

If one observes in the direction Oy ($\theta = 90°$), the ratio $I/\mathscr{E}$ (Lord Rayleigh's ratio) is given by the expression:

$$\frac{I}{\mathscr{E}} = \frac{9\pi^2 N V^2}{2\lambda^4} \left(\frac{n^2 - 1}{n^2 + 2} \right)^2 \tag{9.7}$$

For a given quantity of a material dispersed in unit volume NV = constant, the intensity varies as V. Cloudy media (smoke, suspensions, emulsions, colloidal sols) diffuse much more light than media in which the particles are molecules.

9.10.2   Diffusion by cloudy media

9.10.2.1   Gaseous phase

1) Principle

The photo-chemical decomposition of the vapours of carbon bisulphide in the presence of nitric acid, produce carbon and sulphur particles forming an aerosol. If a parallel beam of light from a carbon arc falls on this medium, a cloud of blue colour develops progressively. If the light diffused in a direction perpendicular to the incident beam is observed through an analyser, it is found to be linearly polarized, the vibration being normal to the plane determined by the light beam and the eye of the observer.

2) Setting up the experiment (fig. 9.31)

a) Apparatus

T: glass tube, 1 m long and 0.1 m in diameter, closed with pyrex plates cemented with picein. It carries two lateral tubules provided with taps R and R'.

M: manometer open to the air.

A: bell-jar containing carbon bisulphide.

B: bottle of nitric acid.

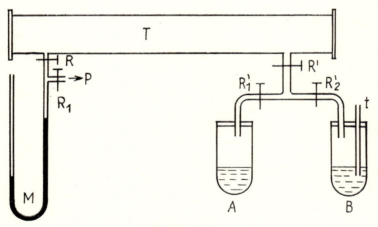

Figure 9.31

b) Adjustments

R' being closed, R and R_1 are opened and vacuum is created at P with the help of a paddle-pump. R_1 now being closed, the manometer M indicates a difference of level of the order of 76 cm. R' is opened and then, with precaution, R'_1; this lets pass into T the vapours of carbon bisulphide produced by the vaporization of the liquid in the container A. When the mercury in the manometer falls to 7 or 8 cm, R'_1 is closed. On opening slowly R'_2 the ambient air is introduced into T; this air having passed through t and the nitric acid in the container B, gets charged with the vapours of the acid. The mercury having fallen by about 30 cm, R'_2 and R' are closed. The experiment is ready.

9.10.2.2 Liquid phase

1) Principle

When sulphuric acid is poured into a solution of sodium thiosulphate, a precipitate of sulphur is formed, the particles growing bigger progressively. At the start of their appearance, these particles satisfy the conditions required for establishing the formula (9.7). Observing in a direction normal to the direction of illumination, the following phenomena are seen to occur progressively: an increase in the diffused intensity (factor V); the change of the bluish colour of the diffused light to white (the law of λ^{-4} is no longer valid when the particles cease to be small in comparison λ); yellow and then red colour of the transmitted light at E, which is more and more deprived of blue.

2) Setting up the experiment (fig. 9.32)

a) Apparatus

   S: carbon arc.
   L: lens of arbitrary focal length.
   C: open glass cell of 5 to 10 cm in length and 4 cm² in section.

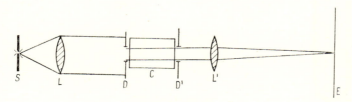

Figure 9.32

D and D': diaphragm with circular apertures.

L': lens with a focal length of 30 cm.

E: white screen.

b) *Adjustments*

Adjustments of S and L: S is at the object focus of the lens *L* which thus gives a parallel beam.

Adjustments of D', L': the lens *L'* forms an image of the circular aperture *D'* on the screen *E* situated 3 to 4 m away.

Preparation of the solution contained in C: C is filled with a $N/4$ solution of sodium thiosulphate $Na_2S_2O_3$. One cm³ of $N/100$ solution of H_2SO_4 is then added and the contents are stirred. Precipitate of sulfur is formed. The speed of evolution is regulated by using a larger or a smaller quantity of H_2SO_4.* The experiment being autocatalytic, it is recommended that the same cell should not be employed twice without rinsing it thoroughly.

3) Variation

Suspensions capable of diffusing light of bluish colour can be obtained in the following ways: in distilled water, put a few drops of a saturated solution of sulphur in acetone or in absolute alcohol (precipitate of sulphur); in ordinary tap water put a few drops of a diluted solution of $AgNO_3$ (precipitate of Ag_2CO_3).

The light diffused by these suspensions in a direction *Oy* perpendicular to the direction *Ox* of the incident beam, is observed through an analyser. It is seen that the transmitted intensity is maximum when the direction of transmission of the analyser is *Oz* and it is nearly zero when the direction of transmission is *Ox*.

On placing a polarizer before *D* (fig. 9.32) and observing in the direction *Oy* without an analyzer, it is seen that the diffused intensity is maximum when the direction of transmission of *P* is parallel to *Oz* and it is almost zero when this direction is parallel to *Oy*.

These results relating to polarization — and which are true only for isotropic particles — show that the assimilation of their induced electric moment to a Hertz dipole is correct. The reader will verify it on the basis of the experiments of § 19.4.2.

---

* For a more detailed description of the phenomena, through suitable only for a long and individual observation, see A. Kastler, "la diffusion de la lumière par les milieux troubles", Hermann, Paris 1952.

9.10.2.3 Diffusion by particles of colloidal gold

It does not follow Rayleigh's law. The gold sol appears as a transparent pink solution, strongly diffusing light of brown colour, which does not follow the simple rules of polarization, the particles even though very small ($\ll 0.1\mu$) being very absorbing.

N.B. In all the preceeding experiments, one should employ distilled water, twice filtered on the millipores membranes, or on a filter of sintered glass $n° 4$, under the action of gravity.

It should be verified that this water and the chemical solutions made with it, show negligible diffusion.

9.10.3 Molecular diffusion

This phenomenon can be observed only individually and in liquid phase alone. The observation in gaseous phase requires very many precautions. The apparatus represented in figure 9.33 is used. It is made of sealed glass.

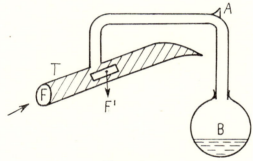

Figure 9.33

T is a cylindrical tube, about 30 cm long and 5 to 6 cm in diameter, closed at one end by a plane window F and the other end being drawn out and curved.

A side opening connects it to a balloon B of capacity 1.5 liters. By means of an opening situated at A, one litre of very pure benzene which has been checked as non-fluorescent, is introduced into B. B is immersed in melting ice and vacuum is created by means of a paddle pump; tube A is then closed with a blow lamp. Benzene is then slowly distilled from B to T by immersing B in water at 50°C and T in melting ice. The liquid is again passed on to B and distillation is repeated. A light beam originating from a carbon arc falls normally on the window F and gets lost in the tapered end, the outside surface of the tube having been blackened with a mat varnish except for a window F' kept for observation (on the face opposite to that shown in the figure). The trace of the beam appears quite bluish. The diffused light is not totally polarized, but the intensity along Ox is almost half of the intensity along Oz: this is because the molecules of benzene are anisotropic.

By illuminating a parallelopipedic block of a good quality, homogeneous artificial crystal, of NaCl for example, it is shown that the diffusion is hardly perceptible, certainly very much less intense than in benzene. In fact, there is almost complete destructive interference of the wavelets diffused transversaly.

10

Reflection, Refraction, Dispersion

10.1 NOTION OF REFLECTING SURFACE

10.1.1 Principle

It is a surface whose inequalities are small compared to the wavelength.
The path difference between a ray of a parallel beam reflected from a hump
of height h and the neighbouring rays reflected from the surface Σ (fig. 10.1)
is $2h \cos i$. Thus the influence of the roughness decreases as incidence increases.

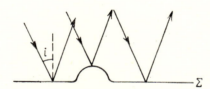

Figure 10.1

10.1.2 Setting up the experiment (fig. 10.2)

a) Apparatus

   S: incandescent lamp.
   D: diaphragm of arbitrary form
   L: lens with a focal length of 15 to 20 cm.
   M: gelatin side of a photographic plate darkened uniformly by exposing
and developing.
   E and E': white screens.

99

b) Adjustments

Distance between S and D: arbitrary.

Adjustment of D, L and E: the lens *L* forms an image of *D*, via the reflecting surface *M*, on the screen *E*.

Adjustment of M: the mirror is set at two positions: the first position, *M*, corresponds to an angle of incidence of 5 to 10 degrees; the image observed on the screen *E* is very hazy. The second position, *M'*, corresponds to an angle of incidence of 70 to 80 degrees; the image observed on the screen *E'* is much sharper.

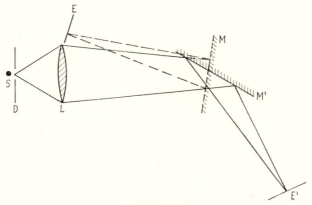

Figure 10.2

10.2   REFLECTIVITY AND REFRACTIVE INDEX

10.2.1   Principle

The reflectivity under normal incidence at the surface of separation between two transparent media of refractive indices (real) n and n', is given by

$$R = \left(\frac{n - n'}{n + n'}\right)^2 \tag{10.1}$$

A piece of glass *G* is immersed in benzene contained in a cell *C*. An image of *G* is formed in red light on the screen *E*. This image is hardly visible since the refractive index of glass is only slightly greater than that of benzene ($n_D = 1.501$). Carbon bisulphide ($n_D = 1.627$) is then added to *C* in small quantities, stirring the mixture after each addition. The glass piece will be rendered invisible when, in accordance with (10.1), the refractive index of the mixture is equal to that of the glass. If white light is used for illumination, the edges of the glass piece reveal their presence by iridescence. This shows that the equality of refractive indices obtained for one wavelength does not hold good for the others (cf. 10.7).

10.2.2   Setting up the experiment (fig. 10.3)

a) Apparatus

   S: carbon arc or incandescent lamp.
   L: lens with a focal length of about 20 cm.
   V: red glass.
   C: cell with parallel faces containing benzene.
   G: piece of glass.
   L': lens with a focal length of about 40 cm.
   E: white screen.

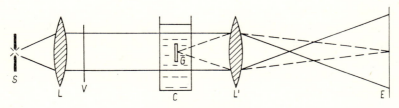

Figure 10.3

b) Adjustments

   Distance between S and L: the lens *L* gives a parallel beam of light.
   Adjustment of C, L' and G: *L'* forms an image of *C* on the screen *E*. The glass *G* is then immersed in the cell *C* and its image is formed on *E* by displacing *L'*.

10.3   REFLECTION FROM ABSORBING MEDIA

10.3.1   Principle

The reflectivity for normal incidence of a substance having a refractive index n and index of absorption $\varkappa$ for a given radiation, is given by the expression:

$$R = \frac{(n - n')^2 + \varkappa^2}{(n + n')^2 + \varkappa^2} \qquad (10.2)$$

Here n' designates the index of refraction of the transparent medium in contact with the substance. R attains high values in the regions of strong absorption; however, since n varies rapidly in these regions, R varies also in a complicated way.

The figure 10.21 shows, for example, the variations of R for cyanine[†] in the visible spectrum.

The light reflected from the surface of a cyanine crystal, fused in contact with glass, is examined by a direct-vision prism. The spectrum of light reflected from glass-cyanine interface, presents a dark band in the green region where the refractive index of the substance is approximately equal to unity and where $\varkappa$ is already small: R is thus small.

10.3.2   Setting up the experiment (fig. 10.5)

a) Apparatus

   S: carbon arc.

   P: crown glass prism with an angle of 5 to 6 degrees (fig. 10.6). On one face of this prism, a cyanine crystal C has been melted by slow heating, which after cooling forms a sticky spot with a glossy surface.

   The light reflected by the surface in contact with air has a purple colour (beam 1, 1′). The light reflected by the surface in contact with glass is green-yellow (beam 2, 2′).

   L: lens of arbitrary focal length.

   L′: lens with a focal length of about 40 cm.

   P′: direct vision prism.

   E: white screen.

b) Adjustments

   Distance between S and L: S is at the focus of L.

   Adjustment of P, L′ and E: L′ forms an image of P on the screen E.

   Adjustment of P′: P′ is placed just after L′.

10.4   COMPARISON OF AIR-GLASS AND AIR-METAL REFLECTION IN NORMAL AND GRAZING INCIDENCES

10.4.1   Principle

The reflectivity of a metallic surface of refractive index n and of index of absorption $\varkappa$, with respect to air, is given by the formula:

$$R_M = \frac{1}{2}\left[\frac{(n - \cos i)^2 + \varkappa^2}{(n + \cos i)^2 + \varkappa^2} + \frac{(n - 1/\cos i)^2 + \varkappa^2}{(n + 1/\cos i)^2 + \varkappa^2}\right] \qquad (10.3)$$

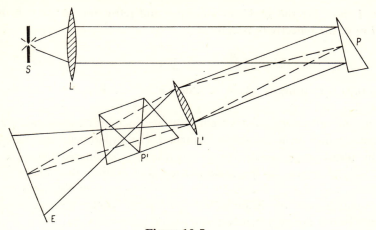

Figure 10.5

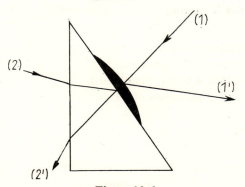

Figure 10.6

where i is the angle of incidence. For $i = 0$ (normal incidence) we get:

$$R_M = \frac{(n-1)^2 + \varkappa^2}{(n+1)^2 + \varkappa^2} = 1 - \frac{4n}{(n+1)^2 + \varkappa^2} \qquad (10.4)$$

For $i = \pi/2$ (grazing incidence), $R_M = 1$.

Thus the reflectivity of a metal changes very little as the incidence is varied from normal to grazing, and is always very high. The reflectivity of a vitreous surface of refractive index n', with respect to air, is given by the formula:

$$R_V = \frac{1}{2}\left[\frac{\sin^2(i-r)}{\sin^2(i+r)} + \frac{\tan^2(i-r)}{\tan^2(i+r)}\right] \qquad (10.5)$$

8*

where *i* and *r* are the angles of incidence and refraction of the light beam. For *i* = 0, the reflectivity becomes:

$$R_V = \frac{(n' - 1)^2}{(n' + 1)^2} \tag{10.6}$$

$R_V = 0.04$ for ordinary glass of index $n' = 1.5$. For $i = \pi/2$, one finds $R_V = 1$.

Thus the reflectivity of a vitreous surface increases considerably as the incidence varies from normal to grazing. The experiment can be carried out by illuminating, under the two incidences considered above, a glass plate half of which has been aluminized. It is shown that under normal incidence, the beam reflected by the aluminized part of the plate is much more intense than the beam reflected by glass. Under grazing incidence, the two beams are very brilliant and are of equal intensity.

10.4.2   Setting up the experiment (fig. 10.7)

a) Apparatus

S *and* S': two carbon arcs.

C *and* C': two ordinary condensers.

L: glass plate half of which is aluminized.

L_1 *and* L_2: two identical lenses of 30 cm focal length.

M: plane mirror.

E: white screen.

b) Adjustments

Distance between S and C: S is very near to the object focus of C.

Distance between S' and C': S' is very near to the object focus of C'.

Adjustment of L: the inclination of L is so adjusted that the beam issuing from C falls on it under grazing incidence and the beam issuing from C' falls under normal incidence. The beam should cover both the aluminized and the non-aluminized parts of the glass plate.

Distance between L and L_1: L is close to the object focus of L_1 and is imaged 4 m away on the screen E.

Adjustment of L, L_2 and M: L_2 forms an image of L on the screen E, the beam having been reflected by the auxiliary mirror M.

The set up of figure 10.7 enables the formation of images E_1 and E_2 side by side on the screen; the image E_1 corresponds to grazing incidence and the image E_2 to almost normal incidence. The comparison can easily be made.

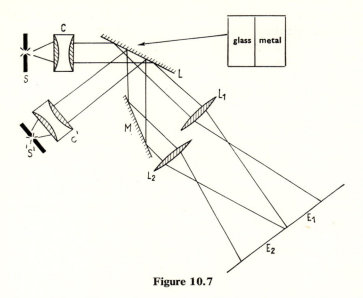

Figure 10.7

10.5 TOTAL REFLECTION AND REFRACTION LIMIT

10.5.1 Principle

When a light ray arrives at a plane surface separating the transparent medium of index n_1 in which it is propagating, from another transparent medium of index n_2 $(n_2 < n_1)$, the ray suffers total reflection if the angle of incidence i is higher than the critical angle l which is given by

$$\sin l = \frac{n_2}{n_1}$$

For the values of i less than l, the reflectivity decreases rapidly (fig. 10.8).

Let us consider the following experiment: two total reflection prisms P and P' made of crown glass are illuminated by a slightly converging beam (fig. 10.9). The hypotenuse faces of the two prisms are placed in contact without any special precautions, such that an air film is formed between them. When the incidence at P is normal, there is total reflection (the angle of incidence on the hypotenuse face is 45° and the critical angle $l \cong 41°$). The trace of the beam at E is bright and on E' it is invisible (fig. 10.10a). By rotating the ensemble of two prisms in the sense of the arrow f, the beam

E' reappears (fig. 10.10c). For an incidence close to *l*, the aspect of figure 10.10b is obtained: the trace of the beam *E'* is separated in two parts, one is completely dark and the other is bright. The trace of the beam *E* is also separated in two parts but their intensities are not very different.

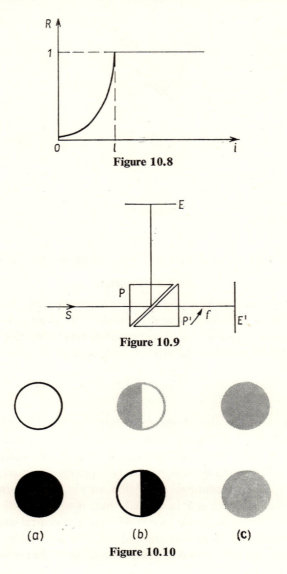

Figure 10.8

Figure 10.9

(a)   (b)   (c)

Figure 10.10

10.5.2   Setting up the experiment (fig. 10.11)

a) Apparatus

S: carbon arc with a red glass or a mercury vapour lamp, A_3 type, with a filter isolating the green line.

C: ordinary condenser.

L: lens of about 30 cm focal length.

D and D': circular apertures.

P and P': two total reflection prisms of crown glass. The ensemble is placed on a platform which can be rotated about a vertical axis.

E and E': white screens.

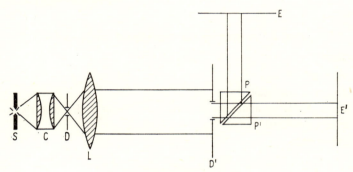

Figure 10.11

b) Adjustments

Adjustment of S, C and D: D is placed at *S'*, the image of *S* given by *C*.

Distance between D and L: D is placed at the object focus of the lens *L* which produces a slightly convergent beam.

Distance between L and D': 3 m.

Distance between D' and P: the aperture *D'* is placed against the prisms and limits the incident beam.

Distance between the prisms: they are in contact.

Adjustment of E and E': E and E' are at 2 to 3 m from the hypotenuse faces of *P* and *P'*.

10.6   LIGHT PIPE

10.6.1   Principle

The light may propagate in transparent solid cylinders if it enters at an incidence suitable for total internal reflections. If a bright point of light is projected, with the help of a strongly convergent optical system, at the entrance of a plastic cylinder, the light beam spreads out in the cylinder and the luminous point is found again at the exit from the cylinder (fig. 10.12).

Figure 10.12

10.6.2   Setting up the experiment (fig. 10.13)

a) Apparatus

S: carbon arc.

C: ordinary condenser.

F: cylinder made of a plastic material, 1 m long and 3 mm in diameter, with carefully polished surfaces.

E: white screen.

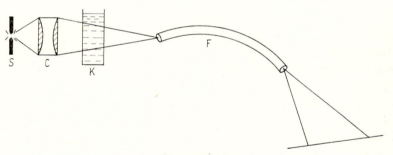

Figure 10.13

b) Adjustments

Distance between S and C: arbitrary.

Distance between C and F: this distance is such that *C* forms an image *S'* of *S* on the entrance face of the tube.

Distance between F and E: the observation screen is at a distance of 2 to 3 m from the exit face of the tube.

c) Comment

A water cell K must be placed between the condenser and the tube so that the entrance face of the tube is not overheated.

10.7   DISPERSION OF REFRACTION

The law of dispersion $n = f(\lambda)$ depends on the substance. This can directly be revealed by the method of crossed spectra.

10.7.1   Method of crossed spectra

10.7.1   Principle

If a parallel beam of light falls on a grating, a spectrum is obtained the linear deviation of which is approximately proportional to the wavelength, the blue having been less deviated then the red. Another grating is placed behind this grating such that the grooves of the two gratings are perpendicular to each other. The deviation produced by the second grating is normal to that of the preceeding one and it is likewise linear in wavelength. The combination of the two gratings thus traces the resultant spectrum, which is approximately a segment of a straight line (fig. 10.14).

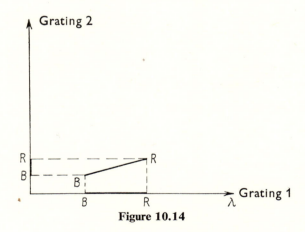

Figure 10.14

A light beam dispersed linearly by a grating is made to fall on a prism of the material under study. The spectrum of the grating will supply the scale of the abscissa (wavelength). The prism is placed such that its edge is perpendicular to the grooves of the grating, i.e. it disperses along the ordinates. (For a transparent prism blue is deviated more than red.)

The resultant of the two orthogonal deviations, represents the dispersion curve of the prism under study (fig. 10.15).

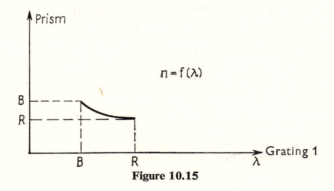

Figure 10.15

The order of succession of the prism and the grating is theoretically unimportant. In fact, it should not be forgotten that the prism must be traversed in planes of the principal section. It is advantageous to place it first. All the rays traverse it almost correctly. Make sure to place the grating very near to the prism and perpendicular to the rays which have suffered a mean deviation (yellow rays for example).

If the prism were traversed by highly divergent rays coming from the grating, the resultant curve of the images will behave in a sense opposite to that of the true phenomenon to be observed.

10.7.1.2   Setting up the experiment (fig. 10.16)

a) Apparatus

   S: carbon arc.
   C: ordinary condenser.
   T: circular source.
   L: lens of 30 cm focal length
   P: prism with a large dispersion (flint).
   R: grating having a small number of grooves.

b) Adjustments

Adjustment of S, C and T: the condenser forms an image of the crater of the arc on the hole *T*.

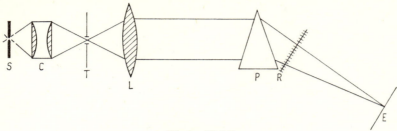

Figure 10.16

Adjustment of T, L and E: the lens L forms an image of T on the screen E about 4 m away.

Adjustment of P and R: the prism and the grating are thus traversed by quasi-parallel beams.

For practical reasons, the prism is placed with its edge vertical. The grooves of the grating are horizontal and the deviation produced is vertical. In fact, the sketch of figure 10.17 is observed.

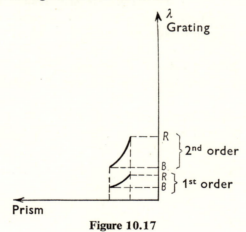

Figure 10.17

10.7.2 Dispersion of refraction in the transparent region

10.7.2.1 Principle (Christiansen's phenomenon)

A transparent solid (in powder form) whose dispersion is represented by the curve S (fig. 10.18) is immersed in a transparent liquid of dispersion represented by L. The equality of refractive indices obtained for a radiation λ_0 does not hold good for others. The reflectivity is zero for the radiation λ_0,

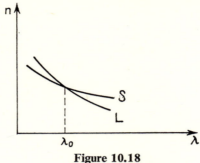

Figure 10.18

which is transmitted whereas the other radiations of the spectrum get reflected a large number of times in all the directions.

The experiment is carried out with the powder of transparent calcium fluoride. The powder ($n_D = 1.433$, a little dispersive) is opaque. Alcohol is poured over it ($n_D = 1.361$). The mixture, quite misty, remains opaque. Benzene ($n_D = 1.501$) is now added in small quantities, stirring the mixture after each addition. At a certain stage, the image S reappears in violet colour. A further addition of benzene changes the colour of the image to red, passing through different colours of the spectrum. The colours are quite pure. Addition of alcohol changes the colour of the transmitted light back to blue.

Dispersion curves, F for calcium fluoride, and L_1, L_2, L_3 for liquid mixtures richer and richer in benzene, are represented in figure 10.19.

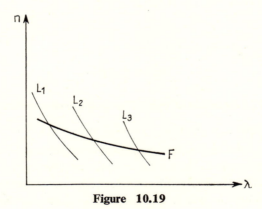

Figure 10.19

Variation

If calcium fluoride is not available, one can proceed as follows: a solution of commercial hydrofluorosilicic acid H_2SiF_6 is poured on solid potassium carbonate till CO_2 ceases to be released.

The refractive index of the precipitate of K_2SiF_6 is very close to that of the solution and the transmitted light is reddish. By the addition of water the colour can be changed to yellow, but the colours are not quite pure and their variations are very much limited.

10.7.2.2 Setting up the experiment (fig. 10.20)

a) Apparatus

  S: lamp having a linear tungsten filament.
  L: lens of 10 to 15 cm focal length.
  C: cell with parallel faces and with a thickness of 1 cm.
  D and D': diaphragms.
  E: white screen.

b) Adjustments

  Adjustment of S, L and E: the lens *L* forms an image of *S* on the screen *E* placed many meters away.
  Adjustments of C, D and D': synthetic calcium fluoride, finely pulverised and sifted through muslin, is introduced into *C* in such a way that the diaphragms *D* and *D'* are covered.

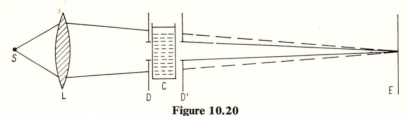

Figure 10.20

10.7.3 Dispersion of refraction near the absorption bands

10.7.3.1 Principle

The absorption determines the dispersion of refraction. The dispersion curves $n = f(\lambda)$ show a rapid variation near the regions of absorption; the index increases on the side of longer wavelengths and values higher than 2 may be attained for the solids; on the side of smaller wavelengths, it may

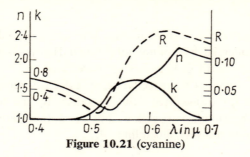

Figure 10.21 (cyanine)

fall below 1 (fig. 10.21). In the region of absorption the index increases as the wavelength increases: the dispersion is said to be "anomalous".

10.7.3.2   Dispersion of sodium vapour in the proximity of D lines

1) Principle

The method of crossed spectra is used, by forming a prism of sodium vapour. The refractive index of this prism departs sensibly from unity only in the proximity of D line, and it is larger at the red end.

   If the auxiliary system disperses horizontally and if the prism of sodium vapour has its base towards bottom, the observed spectrum has the aspect of the figure 10.22.

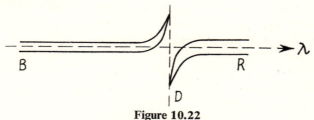

Figure 10.22

2) Setting up the experiment (fig. 10.23)

a) Apparatus
   S: carbon arc.
   L: ordinary condenser.
   F: slit
   B: metal cage containing *M*.
   M: Meker burner.
   L': lens with a focal length of 10 to 15 cm.
   Sp: spectroscope with a flint prism of 60°.
   C: iron cupel containing sodium.

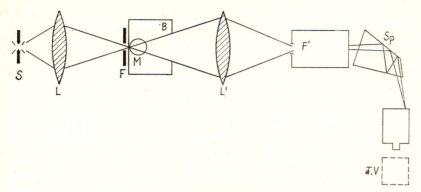

Figure 10.23a

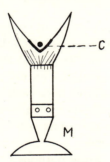

Figure 10.23b

b) Adjustments

Adjustment of S, L and F: the condenser forms an image of the crater *S* on the slit *F* (horizontal).

Adjustment of C: the cupel *C* is placed in the lower portion of the flame.

Adjustment of F and L': *L'* forms an image of *F* on the slit *F'* of the spectroscope *Sp*; *F'* is vertical.

Adjustment of M: the blue conic flame of the burner, charged with sodium vapour, plays approximately the role of a prism with a horizontal edge, and the prism of the spectroscope has its edge vertical. The burner must be placed close to *F* so that the light beam issuing from *F* passes entirely in the upper portion of the flame.

Adjustment of Sp: the spectroscope and the slits *F* and *F'* are so adjusted as to obtain a continuous spectrum of small height. [A piece of sodium is placed in *C*.] As the density of sodium vapour increases in the flame, one observes in the spectroscope, to start with, a dark band in the region of D lines and then a deformation of the spectrum as represented by figure 10.22.

c) Variation

This experiment cannot be projected, but by placing a television camera (*TV*) at the back of the spectroscope eyepiece, suitably drawn out, the figure 10.22 can easily be shown on the screen.

10.7.3.3 Dispersion of cyanine and of fuchsine

A hollow prism of 2 to 3 degrees of angle is constructed with two glass pieces G and G' of 2 cm $\times$ 2 cm (fig. 10.24). Their lower edges, quite straight, are applied against each other and fixed on a glass block S with a cement C insoluble in alcohol. The upper edges are separated by a glass or metal plate of about 1 mm thickness and are held by a rubber band B. Care should be taken that the cement C does not spread to regions close to the edge of the prism. After drying, a drop of a $^1/_{20}$ alcoholic solution of cyanine or of fuchsine is introduced between the plates. The liquid is held due to the capillary action. The prism is placed on the platform of a good Babinet goniometer the slit of which, placed horizontally, is illuminated by the image of a carbon arc (a water tank should be interposed). The parallel beam coming out of the collimator must pass near the edge of the prism so as to minimize the thickness traversed and the strong absorption due to the coloured liquid. Only two narrow portions of the spectrum, one red and the other blue, dispersed vertically, are seen in the telescope. The former is above the latter and is thus more deviated as is evident from the indices of refraction (fig. 10.21). This experiment cannot be projected.

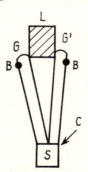

Figure 10.24

11

Polarization by Reflection

11.1 INTRODUCTION

The pencils of rays emitted by most of the common sources have a symmetry of revolution around the direction of propagation. The light is thus said to be *natural*. By treating the natural light in a number of ways (suitable reflections, passage in certain transparent or absorbing crystals) it can lose its symmetry of revolution. The light is said to be *polarized*. The polarization proves that the light vibrations are transversal, i.e. normal to the direction of propagation. In certain important cases, the polarized vibration has a fixed direction: the light is linearly polarized. The instruments which transform natural light into linearly polarized light are known as polarizers; those which can reveal this state of polarization are called analyzers; these are, however, made exactly as the polarizers.

Natural light may be regarded as a linearly polarized light the direction of vibration of which changes at random 10^8 to 10^9 times per second due to the limited duration of wave trains (cf. § 5.1) and consequently does not possess any privileged direction.

11.2 LAW OF MALUS

11.2.1 Principle

Linearly polarized light VV' (fig. 11.1) falls on an analyser which transmits vibrations parallel to the direction OA. Let α be the angle between VV' and OA. Only the component $vv' = VV' \cos \alpha$ traverses the analyser. The intensity I of a light beam (proportional to the square of the amplitude) obtained after passage through the analyser is given by

$$I = I_0 \cos^2 \alpha \qquad (11.1)$$

This is the law of Malus.

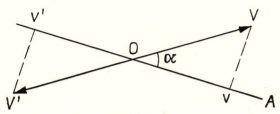

Figure 11.1

11.2.2   Setting up the experiment (fig. 11.2)

a) Apparatus

S: stable source of white light (pointolite lamp or tungsten ribbon lamp of small dimensions).

L: lens of any focal length.

P and *A:* two polarizing sheets. A is mounted on a graduated circle.

R: a photo-cell connected to a projection galvanometer.

D: iris diaphragm.

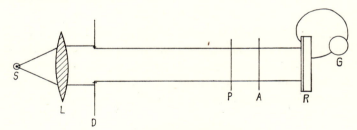

Figure 11.2

b) Adjustments

Distance between S and L: S is at the object focus of *L.*

Distance between L and R: as little as possible (leaving place required by the elements *D, P, A*).

Adjustment of D: the diameter of the circular aperture is such that the light beam covers *P, A* and the sensitive surface of *R*.

Setting of P and A: these are set parallel to each other (maximum deflection of *G*). *A* is turned through 30°, 45°, 60°; the deflection is reduced to 3/4, 1/2, 1/4.

Special precautions: in order to obtain correct results, the experimental set up must be carefully masked against parasite light.

11.3 BREWSTER'S ANGLE

11.3.1 Principle

The ratio of the amplitude a' of the vibration reflected from a surface of separation of two dielectrics, and the amplitude a of the incident vibration (the two vibrations being in the incidence plane) is given by the formula

$$\frac{a'}{a} = \frac{\tan (i - r)}{\tan (i + r)} \tag{11.2}$$

i and r are respectively the angle of incidence and that of refraction. For $\tan (i + r) = \infty$, i.e. $i + r = \pi/2$, a' vanishes and since $n \sin i = n' \sin r$ (n and n' being the refractive indices of the two dielectrics) the angle of incidence i_B, or the angle of Brewster, satisfying the relation (11.2) is such that:

$$\tan i_B = \frac{n'}{n}. \tag{11.3}$$

If natural light is incident at Brewster's angle on the intersurface, the reflected light has no component in the plane of incidence and is thus linearly polarized. The experiment consists in showing that when the light is incident at Brewster's angle on a vertical blackened glass mirror, the reflected light can be extinguished with a polarizer, and that the vibration which it carries is vertical.

One can as clearly demonstrate that when a linearly polarized beam, vibrating in the horizontal direction, falls on this blackened glass mirror, there is no reflected light for $i = i_B$. If the black glass mirror is replaced by a transparent glass, all the incident flux is transmitted when the vibration is horizontal and the incidence is that of Brewster. This can be verified by placing at R a photo-cell and by measuring the galvanometer deflections in presence and in absence of M. Another experiment is set up which demonstrates the variation of Brewster's angle with the refractive index.

9*

11.3.2   Experiment showing that light reflected at Brewster's angle can be extinguished (fig. 11.3)

a) *Apparatus*

   S: incandescent lamp.

   C: ordinary condenser.

   T: circular aperture.

   L: lens with a focal length of 20 to 30 cm.

   M: black glass plate, fixed vertically on the horizontal platform of a Babinet goniometer.

   D: circular aperture.

   A: analyzer, a polarizing sheet.

   E: white screen.

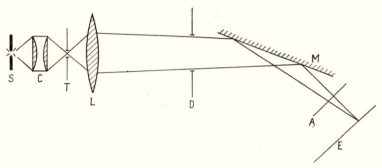

Figure 11.3

b) *Adjustments*

   Adjustment of S, C and T: C forms an image of *S* on *T*.

   Adjustment of T, L and E: lens *L* forms an image of *T* on the screen *E* placed a few meters away. The beam issuing from *L* is thus approximately parallel.

   Adjustment of M: the beam coming from *L* meets the mirror *M* under the incidence of Brewster. To start with the incidence is set to zero by making the beam retrace its path (use the image reflected on the diaphragm *D*) and then the platform of the goniometer is rotated through an angle i_B given by (11.3) measured on the divided circle. For $n = 1.51$, $i_B = 57°$. The angle of incidence may have to be modified by trial and error to obtain complete extinction by the polarizer.

11.3.3 Experiment showing that the light beam reflected under Brewster's angle carries a vibration perpendicular to the plane of incidence (fig. 11.4)

The experimental arrangement is that of figure 11.3. The polarizer P is placed between L and D. The mirror M rotates about a vertical axis and the reflected beam scans the screen E.

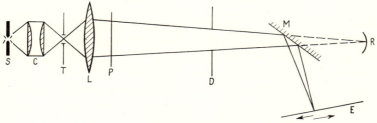

Figure 11.4

11.3.4   Experiment showing the variation of Brewster's angle as a function of the refractive index (fig. 11.5)

a) Apparatus

S: lamp of 4 volts with opalescent glass cover (to be used as a torch).

M_1 *and* M_2: two plane parallel plates, one M_1 of calcium fluoride ($n_1 = 1.43$), the other M_2 of dense flint ($n_2 = 1.70$). Their back surfaces are blackened; in each case, the useful surface is limited to 1 cm² by using black paper masks placed on the upper faces. One can equally well employ two cylindrical cells of 1 cm² cross section. One of these M_1, is filled

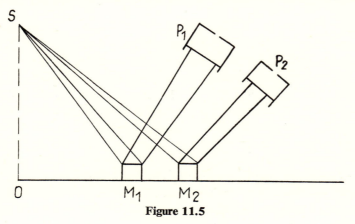

Figure 11.5

with water ($n_1 = 1.33$) and the other M_2 with monobromo-naphthalene ($n_2 = 1.65$). (The difficulty of obtaining a plane surface is larger with a liquid; with solids the difficulty is that of horizontality.)

E: two eye-piece shades, 1 to 2 mm in diameter, which can be rotated about the direction of reflected light. Polarizing pieces P_1 and P_2 are inserted in these shades. The position of the eye is fixed by the shades.

b) Adjustments

Distance between S, M_1 and M_2: the source S is placed 1 m above the plane of the mirrors M_1 and M_2.

Adjustment of S, M_1 and M_2: the distance OM_1 is n_1m, the distance OM_2 is n_2m. SOM_1 and SOM_2 are two vertical planes forming between them a small angle so as not to obstruct the observation on the path of reflected beams.

Adjustment of P_1 and P_2: by rotating P_1 and P_2 about the direction of reflection one can extinguish the reflected beams.

11.3.5   Polarization by reflection on antimony sulphide (fig. 11.6)

a) Apparatus

S: mercury vapour lamp.

D: iris diaphragm.

M: mirror formed of a glass plate covered with a film of antimony sulphide of index $n = 2.7$.

A: analyser, a polarizing sheet.

E: white screen.

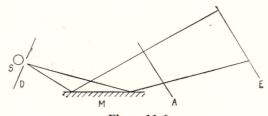

Figure 11.6

b) Adjustments

Distance between S and D: the diaphragm is against the lamp.

Distance between D and M: about 60 cm. The angle of incidence of the light beam is adjusted ($i_B = 70°$) by trial and error by rotating the analyser A

around the reflected beam. A null minimum must be observed on the screen.

Distance between M and E: about 4 meters.

11.4 POLARIZATION BY REFLECTION AIR–WATER

11.4.1 Principle

A beam of natural, white light illuminates a glass cell filled with water under Brewster's angle (fig. 11.7). The beam is split up at the surface of separation air-water. The beam (1) reflected by the surface of separation under Brewster's angle is polarized linearly, the vibration being perpendicular to the plane of incidence. The beam (2) penetrates into the water, is reflected by the bottom of the cell and refracted into the air. Both these beams are received on the screen *E*.

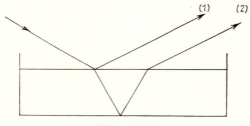

Figure 11.7

If the beam (1) is stopped by means of a polarizer whose direction of transmission is parallel to the plane of incidence, only those rays coming from the bottom of the cell will arrive on the screen *E*. In this way objects lying at the bottom of water can be rendered visible.

11.4.2 Setting up the experiment (fig. 11.8)

a) Apparatus

  S: carbon arc.
  C: ordinary condenser.
  M: plane mirror:
  C': glass cell, with parallel faces, filled with water.
  A: fish made of paper and fixed to the bottom of the cell.
  L: lens with a focal length of 40 cm.
  P: analyser, polarizing sheet.
  E: white screen.

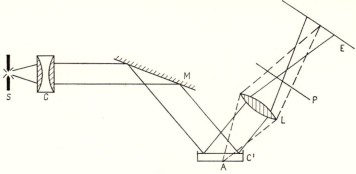

Figure 11.8

b) Adjustments

Distance between S and C: the source *S* is close to the object focus of *C*.

Distance between C and M: about 30 cm.

Adjustment of M: the inclination of the plane mirror is adjusted in such a manner that the reflected beam falls on the tank at an incidence of Brewster ($i_B = 53°$).

Distance between M and C': about 40 cm.

Distance between C' and L: about 50 cm.

Distance between L and E: the lens forms an image of the bottom of the tank on the screen placed about 2 m away.

11.5 ROTATION OF THE AZIMUTH OF A VIBRATION BY REFRACTION THROUGH A GLASS PLATE

11.5.1 Principle

The ratio of the amplitude transmitted at the interface of two dielectrics to the incident amplitude is given by the formula:

$$\frac{b'}{a} = \frac{2 \cos i \sin r}{\sin (i + r) \cos (i - r)} \tag{11.4}$$

The relation holds good when the vibrations are in the plane of incidence. For vibrations normal to the plane of incidence, the ratio is given by:

$$\frac{b''}{a} = \frac{2 \cos i \sin r}{\sin (i + r)} \tag{11.5}$$

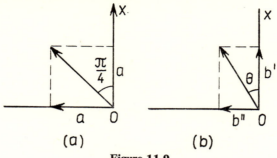

Figure 11.9

Let a linearly polarized light fall obliquely on a dielectric interface (fig. 11.9a), the direction of vibration being at 45° to the plane of incidence (*Ox* is the intersection of the plane of incidence with the plane of the paper) so that its two components are equal. The two transmitted components have unequal amplitudes and the resultant makes an angle θ ($< 45°$) with the plane of incidence. The transmitted vibration has thus been rotated through an angle $\pi/4 - \theta$ with respect to the incident vibration but remains linear.

11.5.2   Setting up the experiment (fig. 11.10)

a) Apparatus

   S: carbon arc.
   C: ordinary condenser.
   T: circular aperture of 1 to 2 cm in diameter
   L: lens with a focal length of 20 to 40 cm
   P: polarizer, polarizing sheet or a nicol
   G: thin plate of non-tempered glass.
   A: analyser:
   E: white screen.

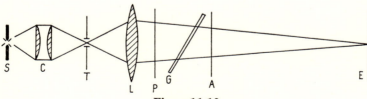

Figure 11.10

b) Adjustments

Adjustment of S, C and T: *C* forms an image of *S* on *T*.

Adjustment of T, L and E: *L* forms an image of *T* on *E* situated many meters away such that the beam coming out of *L* is approximately parallel.

Adjustment of P and A: *P* and *A* are crossed with their directions of transmission at 45° to the horizontal

Adjustment of G: *G* can rotate about a horizontal axis lying in its plane. When it is vertical (*i* = 0) there is extinction on *E*. When *G* is inclined light reappears. The extinction can be re-established by turning *A* in its plane.

11.6   POLARIZATION BY REFRACTION

11.6.1   Principle

The formulae (11.4) and (11.5) show that the ratio of transmitted intensities

$$\varrho = \frac{b''^2}{b'^2} = \cos^2(i - r), \tag{11.6}$$

is equal to 1 for normal incidence, decreases regularly by increasing the incidence but never vanishes. But if the transmission is repeated *N* times, the ratio ϱ is raised to the power *N* and becomes very small. One can thus polarize satisfactorily natural light by making it traverse obliquely a pile of plane parallel glass plates.

This procedure is of particular significance in the infra-red.

11.6.2   Setting up the experiment (fig. 11.11)

a) Apparatus

  S: carbon arc.
  C: ordinary condenser.
  T: circular hole of 1 to 2 cm in diameter.
  L: lens of 20 to 40 cm in diameter.
  G: pile of 5 or 6 plates of non-tempered glass.
  A: analyser.
  E: white screen.

b) *Adjustments*

The same as in 11.5. *A* is removed and *G* is inclined such that the angle of incidence is equal to that of Brewster. The light can almost be extinguished by replacing *A* and rotating it in its own plane.

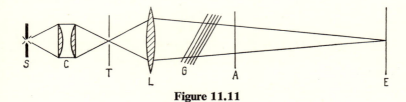

Figure 11.11

12

Polarization by Double Refraction

12.1 DOUBLE REFRACTION BY BIREFRINGENT CRYSTALS

12.1.1 Principle

All crystals except those belonging to the cubic system, are birefringent. One incident ray gives rise to two refracted rays, but these two rays are separated clearly in a small number of crystals only. Calcite (calcium carbonate $CaCO_3$) cleaves easily in the form of a parallelopiped (fig. 12.1). The corner A contains three obtuse angles each equal to 101°53′. If the edges AA', AB and AC are equal, all the faces of the crystal are identical, their form being that of a rhombus; it is a rhombohedron. AA'' is an axis of the ternary symmetry; it makes equal angles with all the faces meeting at A and at A''. This axis is the optic axis of the crystal. The plane containing the optic axis and perpendicular to the entrance face is a principal section of the crystal.

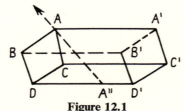

Figure 12.1

A light ray perpendicular to the face $ABCD$ is split into two rays on entering the crystal. The non-deviated ray IO obeys the laws of Descartes*; this is the ordinary ray. The other ray is deviated and does not obey these

---

* Known as Snell's laws in English literature.

128

laws; this is the extraordinary ray (fig. 12.2) *IE*. If the calcite crystal is turned about an axis normal to the entrance face, the extraordinary ray turns around the ordinary ray which remains fixed. There exists an ordinary ray in all those crystals which possess an axis of symmetry of order 3, 4 or 6; they form uniaxial crystals. In the biaxial crystals which have at the most the axes of binary symmetry, neither of the two refracted rays obeys Snell's law. If parallelopiped of iodic acid HIO_3[†] is rotated about the direction *II'* of the ray incident normally, the two rays describe cylinders with axis *II'* (fig. 12.3).

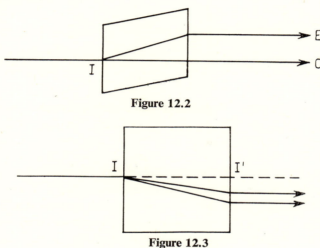

Figure 12.2

Figure 12.3

12.1.2 Experiment showing the existence of double refraction (fig. 12.4)

a) Apparatus

    S: carbon arc.

    C: ordinary condenser.

    D: diaphragm with circular apertures of different diameters.

    B: cleaved crystal rhombohedron of calcite. The experiment can also be conducted with a rhombohedron of sodium nitrate $NaNO_3$[†].

    L: lens with a focal length of 40 cm.

    E: white screen.

b) Adjustments

    Distance between S and C: C forms an image *S'* of *S* on the circular aperture *D*.

Distance between D and O: O forms an image of *D* on the screen *E* placed
3 or 4 m away. In principle, the beam coming out of *O* must be parallel;
it is approximately true when the screen is a few meters away. The diameter
of the circular aperture is so chosen as to obtain a suitable separation of
the two images on the screen.
Distance between O and B: B is quite close to *O*.

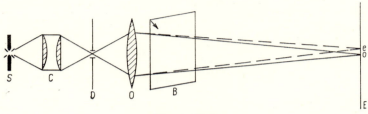

Figure 12.4

12.1.3   Experiment with a biaxial crystal

The same experiment is done with a parallelopiped of iodic acid HIO_3 and
the deviation of the two pencils of rays is observed.

12.1.4   Experiment of Monge (fig. 12.5)

A point source of light seen through a cleaved plate of calcite, gives two
images, the ordinary S_0 on the normal to the plate, and the extraordinary S_e
in the direction $J'S_e$. The path followed by the extraordinary ray must be
thus $SI'J'P (J'P\|SI')$.

If the screen *C* is displaced in the sense of the arrow, the image S_e will
disappear first.

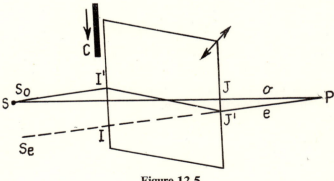

Figure 12.5

12.1.5 Setting up the experiment of Monge (fig. 12.6)

a) *Apparatus*

S: carbon arc.
L_1: lens of about 30 cm focal length.
T: iris diaphragm.
Sp: cleaved plate of calcite, a few cm thick.
L_2: lens of about 30 cm focal length.
E: white screen.
C: black card-board with a straight edge.

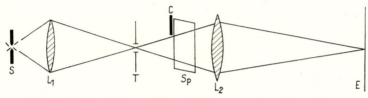

Figure 12.6

b) *Adjustments*

Adjustment of S, L_1 and T: L_1 forms an image of S which covers T.

Adjustment of Sp : Sp is placed normal to the light beam.

Adjustment of T, L_2 and E: L_2 forms the double image of T on E. The diameter of T is such that the two images are separated on E.

Adjustment of C: C placed against the entrance face of Sp is displaced in the direction determined by the centers of the two images.

12.2 THE ORDINARY AND THE EXTRAORDINARY RAYS ARE LINEARLY POLARIZED AT RIGHT ANGLES TO EACH OTHER

12.2.1 Principle

The ordinary and the extraordinary beams coming out of a calcite crystal are linearly polarized at right angles to each other. The extraordinary beam vibrates in the principal section of the crystal and the ordinary beam in the perpendicular plane. To demonstrate this fact the following experiment is carried out. A parallel beam of natural light SI, falls normally on a face of a rhombohedron of calcite (fig. 12.7). On coming out of the crystal, the two beams fall on a black glass plate M under the incidence of Brewster. The crystal is rotated around the direction SI and the beams reflected by the mirror M are observed on a screen.

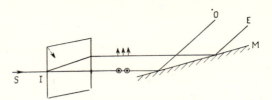

Figure 12.7

The extraordinary image which turns around the ordinary is extinguished when the principal section of the crystal is parallel to the plane of incidence; the ordinary image is then at its maximum of intensity. When the crystal has turned through 90°, one observes the inverse phenomenon: the ordinary image is extinguished and the extraordinary image has its maximum of intensity.

One can as well rotate the mirror (analyser) about the direction *SI*. But then the reflected beams sweep through the space.

12.2.2   Setting up the experiment (fig. 12.8)

a) Apparatus

   S: carbon arc.

   C: ordinary condenser.

   D: diaphragm with circular apertures of different diameters.

   L: lens of 30 cm focal length.

   K: cleaved crystal rhombohedron of calcite or that of sodium nitrate.

   E: white screen.

   M: black glass plate.

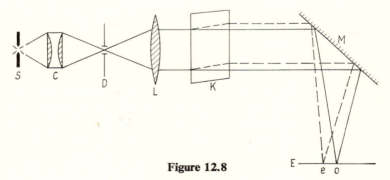

Figure 12.8

b) Adjustments

Distance between S and C: C forms an image S' of S on the aperture D.

Distance between D and L: in principle, the beam emerging from L must be parallel, however it suffices if the image of D is formed on the screen E situated 4 to 5 m away. The diameter of D is chosen, taking into consideration the thickness of K, so as to separate the two images on E.

Distance between L and K: arbitrary.

Distance between K and M: arbitrary. The plate M is so inclined that the light is incident on it at Brewster's angle.

Distance between L and E: 4 to 5 m.

12.3 DOUBLE REFRACTION BY TWO BIREFRINGENT PLATES WITH PARALLEL FACES

12.3.1   Principle

The glass plate M of the experiment 12.2. is replaced by a second calcite crystal K' identical with K. Four images are observed on the screen E (fig. 12.9). The ordinary beam O coming out of K gives rise to two parallel beams on entering K': the beam OO' polarized normal to the principal section of K' and the beam OE polarized parallel to the principal section of K'.

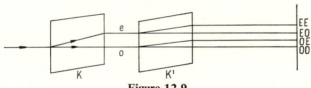

Figure 12.9

Similarly, the extraordinary beam E is split into a beam EO polarized normal to the principal section of K' and a beam EE polarized parallel to the principal section.

When K or K' is rotated about the direction of the incident beam, the images OE, EO and EE rotate around the image OO which remains fixed.

When the principal section of K and K' are parallel, the images OE and EO vanish and the images OO and EE are at their maxima of intensity.

When the principal sections of K and K' are perpendicular, it is the inverse: the images OO and EE vanish and the images EO and OE are at their maxima of intensity.

12.3.2   Setting up the experiment   (fig. 12.10)

The set up is that of figure 12.4; a second rhombohedron of calcite is placed after the first one. The thickness of the two calcite crystals and the diameter of the circular aperture must be suitably chosen so as to obtain four images, well separated, on the screen.

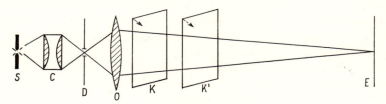

Figure 12.10

12.3.3   Variation

The two plates of calcite may be replaced by two birefringent prisms calcite-glass.

12.4   BIREFRINGENT PRISMS

12.4.1   Principle

The birefringence of uniaxial crystals (calcite and quartz) is studied in certain suitably chosen cases, using the prism method.

Three prisms with equilateral sections are taken. The orientations of their optic axes are as shown on figure 12.11.

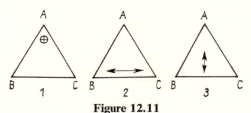

Figure 12.11

The minimum deviations that they produce in a parallel beam of mono-chromatic light are studied. Generally, there are two emerging beams, the ordinary and the extraordinary, which are distinguished by their directions of vibrations (cf. § 12.2.1.) and are not simultaneously at the position of minimum deviation.

In this way, quartz (positive uniaxial $n_e > n_0$) can be distinguished from calcite (negative uniaxial $n_e < n_0$).

Whichever may be the prism and the refracting angle (A, B or C) used for minimum deviation, the deviation is the same for one of the rays (ordinary ray).

In prism 1, the deviations of the two rays are the same whichever refracting angle is used. The surface of indices is a surface of revolution about the optic axis.

In the prism 2, there is only one emergent ray when the refracting angle is A and the position is that of minimum deviation: along the optical axis a uniaxial crystal behaves as an isotropic medium of index n_0. But when B or C is used as the refracting angle, two rays emerge. Their angular deviations are different in the two cases and are less than that obtained in prism 1.

In prism 3, when the refracting angle is A, the deviation of the two rays is the same as in the case of prism 1. If B or C is taken as the refracting angle, the two rays are deviated differently and through an angle smaller than in the case of refraction by A.

12.4.2   Setting up the experiment (fig. 12.12)

a) *Apparatus*

   S: mercury vapour lamp with a green filter.
   C: ordinary condenser.
   F: vertical slit.
   L: lens with a focal length of about 20 cm.
   P: prism with its edge parallel to F.
   A: analyzer (polarizing sheet or nicol).
   E: white screen

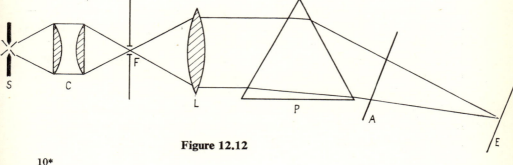

Figure 12.12

10*

b) *Adjustments*

Adjustment of S, C and F: C forms an image of *S* on *F*.

Adjustment of L and E: L forms an image of *F* on the screen *E* placed a few metres away, thus the beam falling on the prism is approximately parallel. The positions of the images on the screen may be measured and this provides an approximate of the deviations.

Adjustment of P: P is placed on a platform which can rotate about its vertical axis and permits the exact replacement of one prism by another.

12.5   NEUTRAL LINES OF A BIREFRINGENT PLATE

12.5.1   Principle

Along all the directions inside a birefringent crystal, two linearly polarized orthogonal vibrations can be propagated independently, having speeds which are generally different. In fact, it is seen that a thin birefringent plate with parallel faces, gives two systems of circular fringes of equal inclination for two incident linear vibrations oriented along two orthogonal directions. These two directions define the two *neutral lines* of the crystal plate.

By rotating a polarizer *P* in its own plane about the beam incident on the plate *L* or about the beam coming out of the plate, two systems of sharp rings are in general observed (they will be hardly distinct if the path differences at the centre differ by one wavelength). They are at their maximum of sharpness for two azimuths of *P*, perpendicular to each other; the transmission direction of *P* is then at 45° to one of the neutral lines. When the transmission direction of *P* is parallel to one of the neutral lines of *L*, only one system of rings is observed.

The existence of the neutral lines can also be demonstrated by illuminating a thin birefringent plate, thickness *e*, placed between two crossed polarizers, by a point source *S* at infinity (fig. 12.13). The emergent beam is received on a screen *E*. It is seen that the interposition of the plate *L* between the crossed polarizers restores the light on the screen *E*.

On rotating the plate in its own plane, two positions are observed, at 90° to each other, for which there is complete extinction on the screen *E*. To these two positions of the plate correspond two orthogonal directions of light vibrations: these are the neutral lines of the plate. The extinction is restored on the screen when the neutral lines are parallel or perpendicular to the transmission direction of *P*.

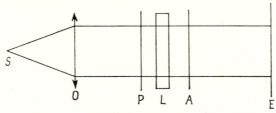

Figure 12.13

12.5.2 Observation of the rings at infinity of a birefringent sheet of mica
(fig. 12.14)

a) Apparatus

S: mercury vapour lamp of medium pressure (type A_2) with a green filter.

C: ordinary condenser.

L: cleaved sheet of mica, 50 to 100 μ thick, the faces of which have been lightly metallized to increase the coefficient of reflection and to obtain high contrast rings by transmission.

P: polarizing sheet.

E: white screen.

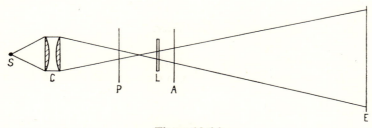

Figure 12.14

b) Adjustments

Adjustment of S, C and L: the condenser converges the light on *L*.

Distance between E and L: a few meters.

12.5.3 Individual observation (fig. 12.15)

a) Apparatus

S: mercury vapour lamp with green filter (type A_2) or sodium lamp (type A_1).

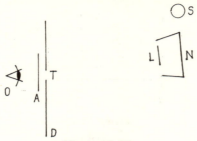

Figure 12.15

L: mica plate without metallization. Rings are observed in reflected light; they have a good contrast (cf. § 3.1). The mica plate is placed in front of a black paper *N*.

P: diffuser with a hole *T* for observation.

A: polarizing sheet to be rotated by the observer *O*.

b) Adjustments

See the figure 12.15.

12.5.4 Observation of a birefringent plate illuminated in parallel light
(fig. 12.16)

a) Apparatus

S: carbon arc with monochromatic filter.

C: ordinary condenser.

D: diaphragm with a circular aperture.

L_1 *and* L_2*:* two identical lenses, of 5 to 6 cm diameter and of 40 cm focal length.

P and A: two polarizing sheets.

L: cellophane sheet or cleaved plate of mica or of gypsum.

E: white screen.

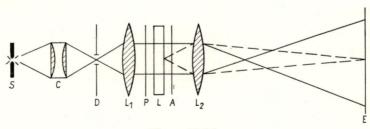

Figure 12.16

b) Adjustments

Distance between S and C: C forms an image *S'* of *S* on the aperture of the diaphragm *D*.

Distance between D and L_1: the circular aperture is at the object focus of L_1.

Adjustment of P and A: the polarizers *P* and *A* are crossed and are placed between the lenses L_1 and L_2.

Adjustment of L_1, L_2 and E: the sheet *L*, placed between *P* and *A*, is a little more than 40 cm from the lens L_2 which forms its image on the screen *E* situated 3 to 4 m away.

13

Polarizers Based on Double Refraction

13.1. INTRODUCTION

The differences in the indices of refraction of ordinary and extraordinary vibrations propagating in transparent uniaxial crystals is put to use for separating the two polarized beams from a beam of natural light.

In birefringent prisms these two beams are conserved. In polarizing prisms only one beam is conserved; the other is eliminated by total reflection.

13.2 BIREFRINGENT PRISMS GIVING LATERAL SEPARATION

13.2.1 Principle

A plane parallel plate of calcite (cleaved rhombohedron) or that of sodium nitrate $NaNO_3$ (natural rhombohedron) of sufficient thickness (1 to 2 cm) permits the separation of two polarized beams provided that the incident beam of natural light is sufficiently narrow. The two beams are parallel. If a polarizing sheet is placed at the exit face of the plate, it can be shown that the two beams are linearly polarized in orthogonal directions.

13.2.2 Setting up the experiment (fig. 13.1)

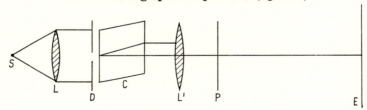

Figure 13.1

141

a) Apparatus

    S: carbon arc with a water tank or a pointolite lamp.

    L: lens of arbitrary focal length.

    D: diaphragm with a circular aperture.

    C: plane parallel plate of calcite or sodium nitrate.

    L': lens with a focal length of 30 cm.

    P: polarizing sheet.

    L'' lens with a focal length of about 2 m

    E: white screen.

b) Adjustments

Distance between S and L: S is at the focus of *L* which produces a parallel beam.

Distance between D and C: D is placed against the crystal plate *C* which receives the beam under normal incidence.

Adjustment of D, L' and E: L' forms an image of *D* on *E*. The circular aperture limits the beam coming from *L*. Its diameter is chosen such that the ordinary beam *O* and the extraordinary beam *E* get separated.

Adjustment of L'': L' is replaced by *L''* and *E* is brought to the image focus. There is only one image of the circular aperture which shows that the *O* and the *E* beams are parallel.

13.3 BIREFRINGENT PRISMS GIVING ANGULAR SEPARATION

13.3.1 Principle

A calcite prism whose refracting edge is parallel to the optical axis (fig. 13.2a) produces an angular separation of the *O* and *E* beams. This element has the inconvenience of deviating the initial beam. The deviation of one of the beams can be compensated by associating with the calcite prism a glass prism *V* (fig. 13.2 b) of the same angle and of the same refractive index for a given wavelength; but in such a calcite-glass birefringent system the undeviated beam is not achromatic since the dispersions of calcite and glass are different (cf. § 10.7.2). This system when employed as a polarizer, must be so placed that the light is incident on the glass prism which is often birefringent (cf. chapter XVI). The deviation of the ordinary beam can finally be compensated and rendered achromatic by using two equal angled prisms of calcite or of quartz cut and associated in the manner shown in figure 13.2 c: this is Rochon's prism.

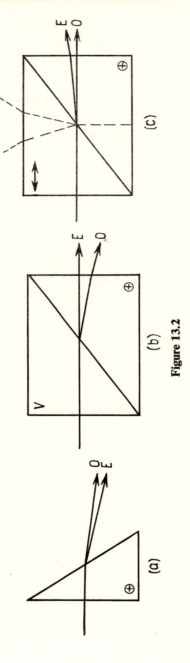

Figure 13.2

If light traverses this prism not from left to right, but from bottom upwards, one obtains the prism of Wollaston which deviates the two polarized
beams through sensibly equal angles on either side of the incident direction
(rays shown by dotted lines in figure 13.2c).

13.3.2   Setting up the experiment

The arrangement of the figure 13.1. is set up, the crystal plate having been
replaced by diverse birefringent systems. The properties described above
are verified. In particular, there always exit two images in the focal plane
of the lens L''. The chromatism of the beams manifests itself by very clear
red and blue iridescence on the edges of the images.

13.4   TOTAL INTERNAL REFLECTION POLARIZERS

13.4.1   Principle

These are made from a parallelopiped of calcite (or of sodium nitrate) cut
into two halves and then cemented together with a thin dielectric in between.
The ordinary beam, having the higher refractive index, gets totally reflected
on the surface of the dielectric film. The angle of incidence of this beam
must be at least equal to the critical angle l_0 so that the polarizer can function; on the other hand if the angle of incidence attains the critical value l_e
relative to the extraordinary beam, this one is also totally reflected and no
light passes through the polarizer.

Thus the convergence of the beam admitted by such a polarizer has two
limits, which define its field.

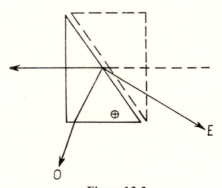

Figure 13.3

The diverse polarizers differ from one another by the orientation of the parallelopiped with respect to the optic axis and by the nature of the dielectric. In the polarizers of Foucault and of Glan, it is simply an air film. In the polarizers of Nicol and of Glazebrook, commonly employed, it is a cement with an index lying between n_0 and n_e.

If one has a prism with air film, one can, by placing one of the two halves at C in the arrangement of figure 13.1, show that the beam O is totally reflected at the surface of separation (fig. 13.3 is made for a Glan), the beam E is deviated and that both the beams are strongly iridescent. On placing the second half also (shown dotted in fig. 13.3), the beam E, undeviated and achromatic arrives at the screen E'.

The field of any polarizer can be determined by placing it at C (fig. 13.1). A slightly converging beam issuing from D is incident normally on the face AB (fig. 13.4) of the polarizer [position (1)]. Then the polarizer is inclined with

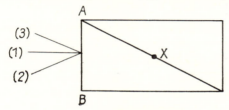

Figure 13.4

respect to the incident beam, by rotating it about an axis normal to the figure at the point X and situated in the plane of the cut. In position (2), one observes on the screen a region in which the light is natural since none of the beams has suffered total reflection. This region is separated from the region where the beam O has suffered total reflection by a system of interference fringes and a red band R (fig. 13.5a). In the position (3), one observes a dark region (both the beams have been totally reflected) separated from the region of polarized light by a blue band (fig. 13.5b).

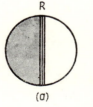

Figure 13.5

The angular interval (2)–(3) is of the order of 6 to 8 degrees for the Foucault and the Glan, 30° for Nicol cemented with Canada balsam and a few degrees in most of the Glazebrooks cemented with balsam.

13.4.2   Setting up the experiment

The experimental arrangement is that of figure 13.1 in which the component C is replace by diverse polarizing prisms.

13.5   SPECIAL BIREFRINGENT PLATES

13.5.1   Principle

Let us consider the set-up of figure 12.13. Let OP be the direction of the linear vibration transmitted by the polarizer, OX and OY be its projections on the neutral lines Ox and Oy of the birefringent plate L and α be the angle between OP and Ox (fig. 13.6).

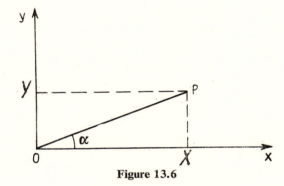

Figure 13.6

On coming out of this plate of thickness e, the phase difference between OX and OY is:

$$\varphi = (2\pi/\lambda)\,(n_x - n_y)\,e \tag{13.1}$$

a) for arbitrary value of φ, the resultant of OX and OY gives an elliptic vibration.

b) if L is a quarter wave plate (or an odd number of times a quarter-wave)

$$(n_x - n_y)\,e = (2K + 1)(\lambda/4) \quad K \text{ being a whole number} \tag{13.2}$$

The elliptic vibration has for axes Ox and Oy. If, in addition, $\alpha = \pi/4$, the resultant vibration is circular.

c) If L is a half-wave plate (or odd number of times a half-wave)

$$(n_x - n_y)e = (2K + 1)(\lambda/2)   K \text{ being a whole number}   (13.3)$$

the resultant vibration is linear, symmetrical to OP with respect to the neutral lines of the plate. If the polarizer and the analyser were crossed in the absence of the plate, the extinction can be restored by rotating the analyser through 2α.

d) If L is a full-wave plate (or a whole number of times a wave)

$$(n_x - n_y)e = K\lambda   K \text{ being a whole number}   (13.4)$$

the emerging vibration is linear and parallel to OP.

13.5.2 Setting up the experiment (fig. 13.7)

a) Apparatus

$S:$ mercury vapour lamp with a green filter.

$C:$ ordinary condenser.

$D:$ diaphragm with a circular aperture.

P and A: two polarizing sheets.

$O:$ lens of 30 cm focal length.

L and L' : quarter-wave plate (L) and half-wave plate (L'). It is not easy to obtain achromatic quarter-wave and half-wave plates. Using mica sheets we make the wave plates corresponding to the green line of mercury. When used in white light, the residual colours will appear and show the chromatism of the mica. The wave plates are mounted in frames.

$E:$ white screen.

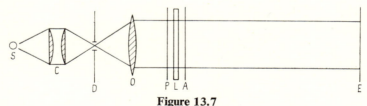

Figure 13.7

b) Adjustment

Distance between S and C: C forms an image of S on the aperture of D.

Distance between D and O: the circular aperture is at the object focus of O.

Adjustment of P, A and L: the two polarizers are placed after the lens O with the plate L between the two. In order that the neutral lines of L are at 45° to the vibration transmitted by P, L is rotated in its own plane till the image intensity on the screen E remains unmodified by a rotation of the analyser about the emerging beam. Or L may be fixed on a roughly graduated circle; L is rotated till there is extinction between crossed P and A, it is then further rotated through 45° with the help of the graduated circle.

13.5.3   Double-plate of Bravais

This is formed by the juxtaposition of two full-wave plates corresponding to the mean yellow, their neutral lines being crossed. It is obtained as follows: a mica plate is cut along a direction making an angle of 45° with the neutral lines; one of the two halves is rotated such that the two faces interchange their positions; the two halves are then brought close to each other along the line of cutting.

Between crossed polarizers, the plate presents the sensitive tint (purple, complementary of the mean yellow) over all its surface. If an elliptical vibration falls on this plate, the phase difference between the two components of the vibration resolved parallel to the neutral lines of the plate, is added to the phase difference of the plate in one half of the double plate and is substracted in the other half, the hue changes towards red in one and towards blue in the other. The double-plate, thus, forms a polariscope for elliptical vibrations.

13.6   FRESNEL'S PARALLELOPIPED

13.6.1   Principle

This device plays the role of an achromatic quarter-wave plate. It is based on the fact that the phase difference between the incident and the reflected vibrations accompanying total internal reflection (air-glass) is not the same when the incident vibration is parallel or perpendicular to the plane of incidence. For a certain angle of incidence (53° for a glass of index 1.5) the phase difference between parallel and perpendicular vibrations is 45°. A linearly polarized ray SI (fig. 13.8) vibrating at 45° to the plane of the figure falls normally at the face AB of a glass parallelopiped $ABCD$. The angle A is 53° and this is also the angle of incidence at I. The two equal components of the linear vibration acquire a phase difference of 45° which is doubled by the second reflection under the same angle at I'. A circular vibration emerges from the parallelopiped. The intersection of the plane of two reflections with the entrance face acts as the neutral line with smaller index of refraction of the equivalent quarter-wave plate.

By using two parallelopipeds disposed in the manner shown in figure 13.9, one obtains the equivalent of a half-wave plate and the lateral displacement of the light beam is eliminated.

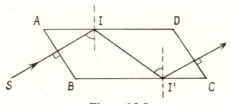

Figure 13.8

13.6.2 Setting up the experiment

The scheme is that of figure 13.7, it is shown that the rotation of A does not bring about any variation of intensity of the beam (in case of a single parallelopiped). S may be a white light source.

13.7 CIRCULAR POLARIZERS

A circular polarizer transforms natural light into circularly polarized light. It consists of a linear polarizer followed by a quarter-wave plate or a Fresnel's parallelopiped.

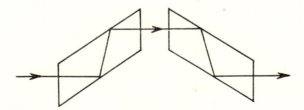

Figure 13.9

The direction of vibration OP furnished by the linear polarizer makes an angle $\alpha = \pi/4$ with the neutral lines of the plate or with the intersection of the symmetry plane of the parallelopiped with the entrance face. The sense of the circular vibration (for an observer receiving the beam) is the one along which the vibration OP has to be rotated through $\pi/2$ to bring it on to the neutral line of higher index (also known as slow axis). This neutral line of a plate can be determined by the method described in appendix C;

for Fresnel's parallelopiped this is the direction perpendicular to the plane of the figure 13.8. A right handed circular vibration turns clockwise for the observer.

A circular analyser consists of a quarter-wave plate followed by a linear polarizer with its transmission direction at 45° to the neutral lines of the plate. A right (or left) circular polarizer traversed in opposite sense by the light, stops the right-handed (or left-handed) circular vibrations.

14

Interference in Polarized Light

14.1 CHROMATIC POLARIZATION

14.1.1 Principle

Consider a plane parallel birefringent plate illuminated by linearly polarized light. The incident vibration may be broken down into two coherent vibrations directed along the neutral lines of the plate. On coming out of the plate, these vibrations have a phase difference, φ, and can interfere if rendered parallel with a polarizer.

The phase difference φ varies continuously from one end of the spectrum to the other, and one observes on the screen a light beam in which the different radiations are unequally affected. When one or two monochromatic radiations are completely eliminated by interference, the superposition of other radiations gives rise to colour on the screen.

If the thickness of the plate is not constant, one observes different interference colours by bringing the plate in focus. By rotating the plate in its own plane, the colours have maximum purity when the neutral lines of the plate are at 45° to the incident vibration *OP*.

If the analyser is rotated in its own plane, the colours observed with polarizers parallel are complementary to the colours with polarizers crossed.

14.1.2 Setting up the experiment (fig. 14.1)

a) *Apparatus*

   S: carbon arc.

   C: ordinary condenser.

   P and A: two polarizing sheets.

L: birefringent plates of different thicknesses giving different colours. One can equally well observe the crystallization of sodium thiosulphate* placed on the object slide of a microscope. One can also obtain coloured pictures using cellophane sheets which have been stuck together.

O: lens of 15 cm focal length.

E: white screen.

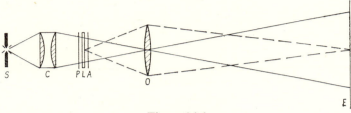

Figure 14.1

b) Adjustments

Distance between S and C: the image of *S* is formed at about 30 cm from *C* on the lens *O*.

Adjustment of P, A and L: *P* is placed against the condenser and *L* at about 15 cm from *C*. The position of the analyser is unimportant provided that whole of the beam passing through *L* passes through the analyser too.

Adjustment of L, O and E: the lens *O* forms an image of the plate *L* on the screen *E* situated 3 to 4 m away.

14.1.3   Observation of a large birefringent plate (fig. 14.2)

The following experiment is very suitable for projecting large images (about 2 m in diameter).

a) Apparatus

S: carbon arc.

C: ordinary condenser.

m: plane mirror.

M: concave spherical mirror with a diameter of 15 to 20 cm and with a radius of curvature of 50 cm.

P and A: two polarizing sheets.

---

* By raising the temperature very slightly, some crystals of sodium thiosulphate are dissolved in their water of crystallization placed between two glass plates which are held above a small flame. The liquid film is then left to cool.

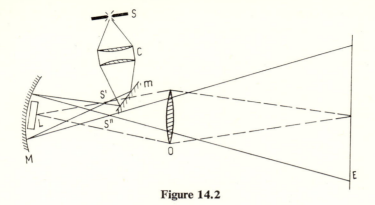

Figure 14.2

L: colourless cellophane sheets of large dimensions (15 to 20 cm in diameter).

O: lens of 40 cm focal length.

E: white screen.

b) Adjustments

Adjustment of S, C and m: the distance between *S* and *C* and the inclination of *m* are so adjusted that the image *S'* of *S* given by *C* is formed approximately at the centre of curvature of the spherical mirror *M* after reflection of the light beam from *m*.

Adjustment of M: the mirror *M* is so inclined that the image *S"* of *S'* given by *M* is by the side of *S'*.

Adjustment of L and O: *L* is placed against the spherical mirror and the lens *O* forms its image on the screen *E* situated 3 or 4 m (or more) away.

14.2 SUPERPOSITION OF TWO HUES GIVEN BY A BIREFRINGENT PLATE BETWEEN TWO CROSSED AND PARALLEL POLARIZERS

14.2.1 Principle

A point-source of white light illuminates a birefringent prism *B*, which produces lateral separation, and the emerging beam falls on a screen *E* (fig. 14.3). The lateral separation given by *B* is so adjusted that the *o* and *e* images, polarized in perpendicular directions, overlap partially on the screen *E*. With a polarizer *P*, one of the two images, *e* for example, is extinguished. For the *e* image, the birefringent prism acts as a polarizer crossed with *P*. For the *O* image, *B* acts as a polarizer parallel with *P*.

A mica sheet or a colourless sheet of cellophane set at 45° to the principal section of *B*, is now interposed between *B* and *P*. The two beams, ordinary and extraordinary, give rise to interference phenomena. Two coloured spots are observed on the screen. One beam (*o*) shows the phenomena between parallel polarizers and the other (*e*) between crossed polarizers. The hues are thus complementary and in the region common to (*o*) and (*e*), the superposition of the two hues gives white light.

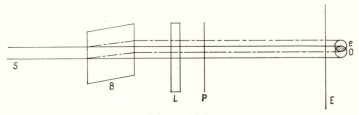

Figure 14.3

By folding *N* times, the cellophane sheet in the same sense, parallel to the neutral line, its thickness becomes *N* times more and one observes the variation of hues as a function of the thickness.

14.2.2   Setting up the experiment with a calcite rhombohedron (fig. 14.4)

a) *Apparatus*

S: carbon arc.

C: ordinary condenser

D: diaphragm with a circular aperture of variable diameter.

B: cleaved rhombohedron of calcite giving a sufficient separation (thickness of the order of 5 cm).

L: colourless sheet of cellophane.

O: lens with a focal length of 30 cm.

A: analyser, a polarizing sheet.

E: white screen.

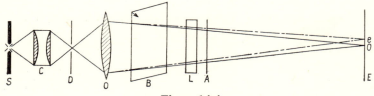

Figure 14.4

b) Adjustments

Distance between S and C: arbitrary, provided that the condenser forms the image of the source on the circular aperture.

Distance between D and O: O forms an image of the circular aperture on the screen *E* placed 3 to 4 m away.

Distance between O and B: arbitrary.

Distance between B and L: arbitrary.

Adjustment of D: the diameter of the aperture is so adjusted that the two images observed on the screen partly overlap.

14.2.3 Setting up the experiment with a calcite-glass prism (fig. 14.5)

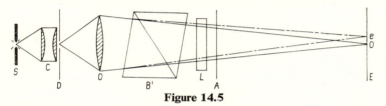

Figure 14.5

a) Apparatus

All the elements shown in figure 14.4. are used; the calcite rhombohedron *B* is replaced by a calcite-glass prism *B'* of 7 to 8 cm thickness.

b) Adjustments

Distance between S and C: C forms an image of *S* at a distance of about 35 cm.

Distance between C and D: D is placed against the condenser.

Distance between D and O': O is placed at the image of *S* to obtain maximum of light on the screen *E*; *O* forms an image of the circular aperture on the screen *E*, 3 to 4 m away.

Adjustment of B' and the diameter of the aperture: the diameter of the circular aperture and the distance *B'D* are so adjusted as to obtain on the screen two images overlapping by one half.

14.3 BABINET COMPENSATOR, BIOT'S BOWL

14.3.1 Principle of Babinet Compensator

A Babinet compensator consists of two quartz wedges of equal angle θ, the optic axes being oriented as shown in figure 14.6. A light beam traverses the system normal to the faces of the plate, i.e. normal to the axes.

One of the two wedges can be displaced by means of a micrometer screw: for example, the wedge $AA'B'CC'D'$ may slide in the direction of its optical axis remaining at the same time in contact with the other wedge.

Between polarizers, one observes straight equidistant fringes parallel to the edge of the wedge and localised inside the system.

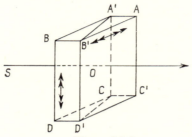

Figure 14.6

At a distance x from the point where the two wedges have equal thickness (fig. 14.7) the path difference between the ordinary-extraordinary ray and the extraordinary-ordinary ray is given by:

$$\delta = 2(n_e - n_0)\,x \tan \theta \qquad\qquad (14.1)$$

The fringes have maximum contrast when the direction of transmission of the polarizer and the analyser are at 45° to the neutral lines of the compensator.

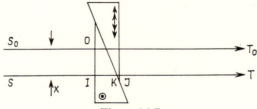

Figure 14.7

In white light, coloured fringes are observed. The central fringe corresponding to $x = O$ is dark when the polarizers are crossed, and white when the polarizers are parallel. When the moveable wedge is displaced, the fringe system gets displaced without any change in the aspect.

One can determine the path difference introduced by a crystal plate C placed before or after the compensator and whose neutral lines are parallel to those of the compensator. It suffices to determine the displacement of the fringes of the compensator on the introduction of the crystal plate, by focussing the cross-wires on the compensator.

In the experiment to be described, the fringes of the compensator are observed in monochromatic light. The plate to be studied is a cellophane sheet which is introduced in one half of the light beam. One observes the fringe displacement in that part of the field of view where the light has traversed the cellophane.

The path difference given by the cellophane sheet is determined by the distance between the two dark fringes belonging to the fixed and the displaced fringe system. In this measurement, there is an indeterminacy equal to an integral multiple of wavelength, due to the fact that the central fringe cannot be distinguished from the others in monochromatic light. To remove this indeterminacy, the experiment is repeated with white light and the displacement of the central fringe is measured.

14.3.2 Observation of the fringes of Babinet's compensator (fig. 14.8)

a) Apparatus

S: carbon arc with a filter for observation in monochromatic light.
C: ordinary condenser.
P and A: polarizers, two polarizing sheets.

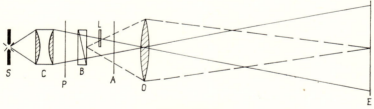

Figure 14.8

B: Babinet compensator.
L: cellophane sheet.
O: lens of 30 cm focal length.
E: white screen.

b) Adjustment

Distance between S and C: this distance is such that the condenser forms an image of the source at a distance of about 60 cm.

Adjustment of P, A, B and L: the polarizers *P* and *A* are crossed; the neutral lines of *B* and of *L* are parallel to each other and are at 45° to the transmission directions of *P* and *A* (fringes of maximum contrast).

Distance between C and P: the lens *O* is at the image of *S*. It forms an image of the Babinet compensator on the screen *E*, 3 to 4 m away.

14.3.3 Principle of Biot's bowl

Biot's bowl consists of a plane parallel plate and a plano-concave lens (fig. 14.9). These elements are made of a uniaxial crystal with the optical axes parallel to the surfaces and crossed with each other as indicated in the figure 14.9. The thickness of the plane parallel plate is equal to the thickness of the lens at its centre.

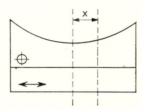

Figure 14.9

Between two polarizers, localised in the device, a system of concentric rings are observed centred at the point where the thicknesses of the two constituent elements are equal. At a distance *x* from this point, the path difference between the ordinary-extraordinary ray and the extraordinary-ordinary ray is given by:

$$\delta = (n_o - n_e)\, x^2/2r \qquad (14.2)$$

where *r* the radius of curvature of the lens.

14.3.4 Observation of the rings given by a Biot's bowl

The experimental set up is that of figure 14.8 in which the sheet of cellophane and the Babinet compensator are replaced by a Biot's lens.

14.4 CHANNELED SPECTRUM PRODUCED BY A BIREFRINGENT PLATE

14.4.1 Principle

In the set-up of the figure 14.1, let the plate be sufficiently thick so that the condition $(n'' - n')\, e = K\lambda$ (*K* being a whole number), holds good for at least 4 or 5 visible radiations. The radiations extinguished by the optical system are evenly distributed throughout the spectrum and the hue obtained on the screen appears white. This is the white of higher order, which when analysed with a spectroscope, gives a channelled spectrum.

In order that the dark bands be completely black and the bright ones at their maximum of intensity, the polarizers must be either crossed or parallel, and the neutral lines of the plate must be at 45° to the transmission directions of the polarizers P and A.

When the polarizers are changed from crossed position to parallel position, the dark bands change to bright ones, and vice versa.

14.4.2   Setting up the experiment (fig. 14.10)

a) Apparatus

S: carbon arc.

C: ordinary condenser.

F: vertical slit of variable width.

P *and* A: polarizers, two polarizing sheets.

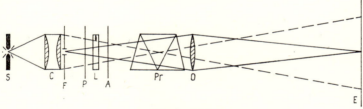

Figure 14.10

L: quartz plate cut parallel to the optic axis and at least 1 mm thick.

P_1: direct vision prism.

O: lens of 40 cm focal length.

E: white screen.

b) Adjustments

Distance between S and C: this distance is such that C forms an image of S at a distance of 45 cm, inside the direct-vision prism.

Distance between C and F: the slit is against the condenser.

Distance between F and O and between O and E: the lens O, placed just after Pr, forms an image of F on the screen E, 3 to 4 m away.

Adjustments of L and of P and A: the polarizers and the plate must lie between the slit and the direct-vision prism. They must be traversed by whole of the beam. In principle, L must be illuminated by a parallel beam. This is practically achieved if the condenser forms an image of the source sufficiently far away.

c) Observation

If the analyser A is replaced by a birefringent prism, two channelled spectra are observed on the screen, the whites of one spectrum coinciding with the blacks of the other.

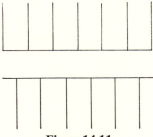

Figure 14.11

14.5   OBSERVATION OF CRYSTAL PLATES IN CONVERGENT LIGHT

14.5.1   Principle (fig. 14.12)

The crystal plate L, placed between polarizers P and A is illuminated by a parallel beam incident at a certain angle. The emergent beam converges to a point in the focal plane of the lens O. The two plane wavefronts produced by the plate interfere and the state of interference at point M is characteristic of the angle of incidence of the parallel beam on the plate. If the plate L is illuminated in convergent light, one observes on the screen E the fringes corresponding to all the inclinations.

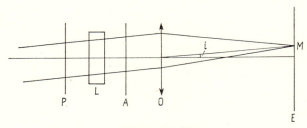

Figure 14.12

A uniaxial crystal plate cut parallel to the optic axis is observed in convergent light: the fringes appear only when the light is monochromatic; these are hyperbolas with axes parallel and perpendicular to the optical axis. The centre of the hyperbolas is dark or bright depending on whether the polarizers are crossed or parallel. By associating two identical plates with their axes crossed, the hyperbolas become visible in white light.

If the uniaxial crystal (calcite for example) is cut perpendicular to the optical axis, one observes, in white light, coloured rings centered on the trace of the optical axis and a dark (crossed polarizer) or a white (parallel polarizers) cross. The two branches of the cross are the diameters of the interference pattern and are parallel and perpendicular to the transmission directions of the polarizers. This cross is not visible in the central region when a quartz plate cut perpendicular to the optical axis is observed in a convergent beam of white light. The rays arriving at the centre of the field traverse the quartz plate in the normal direction and are, therefore, affected by the rotatory power of quartz (cf. § 17.2).

Biaxial crystals show much more complicated figures of interference.

14.5.2 Setting up the experiment (fig. 14.13)

a) *Apparatus*

S: carbon arc.

C and C': two identical condensers.

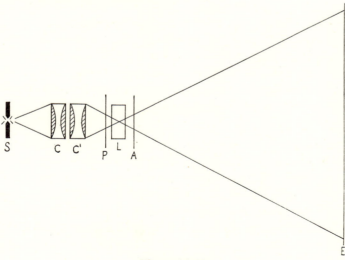

Figure 14.13

L: crystal plates to be studied.

P and A: polarizers, two polarizing sheets.

E: white screen.

b) *Adjustments*

Distance between S and CC': the two condensers are close together and are as far away from the source as possible so as to give a highly converging beam.

Distance between C' and L: L is placed at the point of convergence of the beam.

Distance between L and E: 3 to 4 m, which is practically infinite and thus the observation in the focal plane of a lens can be avoided.

14.6 POLARIZATION INTERFEROMETERS

14.6.1 Principle of interferential contrast

A point source of white light, situated at infinity, illuminates a phase object *A* having a defect of optical thickness in a very small region *M*. The lens *O* forms an image of *A* on the screen of observation *E* (fig. 14.14).

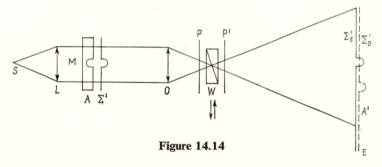

Figure 14.14

The plane wavefront Σ' emerging from *A* is deformed due to the phase defect at *M* (cf. § 9.7.1). By means of a special Wollaston prism *W* placed between polarizers, Σ' is split into two wavefronts Σ'_0 and Σ'_E sheared laterally; these are represented in the image plane *A'*. These are polarized in perpendicular planes and may interfere with the help of an analyser.

On emerging from the plate, Σ'_0 and Σ'_E have a longitudinal displacement due to the path difference introduced by *W*, and a lateral displacement due to the birefringence of *W*.

The Wollaston W is so chosen that the lateral displacement is small in comparison to the lateral dimensions of the defect M. Let the path difference δ between Σ'_0 and Σ'_E be equal to $\lambda/2$ for the mean yellow light.

In the regions where the wavefront is plane (fig. 14.15), there is a purple hue. In the deformed regions, the hue changes revealing the "slopes" of the phase object which is thus rendered visible. This method brings to light the gradient of the optical thickness of the object.

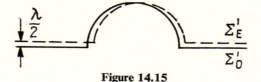

Figure 14.15

By displacing the Wollaston W in the direction of the two arrows (fig. 14.14) δ varies and one may obtain all the colours of Newton's scale between crossed or parallel polarizers.

The birefringent system used in the experiment is a large field Wollaston prism. The convection currents produced by a candle flame are shown. The phase object consists of the variations of the refractive index of air due to variations of temperature.

14.6.2 Principle of the large field Wollaston prism

The isochromatic lines of a Wollaston prism (cf. § 13.3.1) illuminated in a convergent beam (cf. § 14.5.1) and placed between crossed polarizers, are the hyperbolas which limit the field. The system used (fig. 14.16) con-

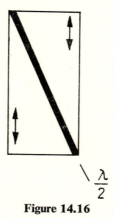

Figure 14.16

sists of two quartz wedges with optical axes parallel, separated by a half-wave plate oriented at 45° to these axes. In this way a rather more converging beam (more than 20° of angle) can be used.

14.6.3   Observation of interference fringes with a large field Wollaston prism

Consider the set up of figure 14.2. in which the lens O forms an image of the spherical mirror M on the observing screen E. The Wollaston prism W is placed at the centre of curvature of the mirror M (fig. 14.17).

An incident ray (1) (fig. 14.18) is split into two rays $(1)_E$ and $(1)_0$) on traversing the Wollaston; these get reflected from the mirror M and again traverse the Wollaston in opposite direction. On emerging from the Wollaston the two rays are combined, and there is now one single ray carrying the ordinary and extraordinary vibrations. If the axis of the Wollaston and that of the mirror coincide, the ordinary and the extraordinary rays are in phase and, with parallel polarizers, the screen is uniformly illuminated in white light.

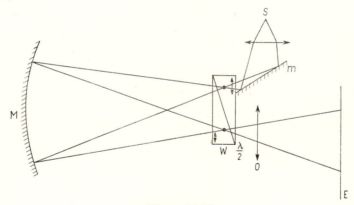

Figure 14.17

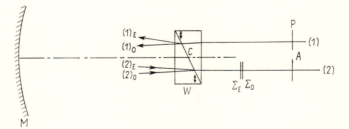

Figure 14.18

By displacing the Wollaston perpendicular to the mirror axis, a path difference δ is introduced between the ordinary and the extraordinary rays. The uniform colour observed on the screen depends on the value δ. If δ is large, one obtains the white of higher order.

If W is no longer at the centre of curvature of the mirror (fig. 14.19), two rays $(2)_0$ and $(2)_E$ are obtained which are no longer parallel and which have a certain path difference between them. One observes, on the screen, a system of straight fringes analogous to Young's fringes and the fringe

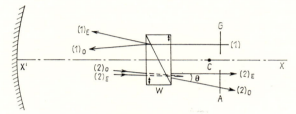

Figure 14.19

spacing depends or the angle θ between the rays considered. The angle θ depends on the distance between C and the Wollaston: the fringe spacing can thus be varied by displacing the Wollaston parallel to the axis of the mirror.

14.6.4   Observation of a channelled spectrum (fig. 14.20)

The experimental set up is that of figure 14.17. The light beam coming out of the analyzer falls on an auxiliary plane mirror m which reflects it on to a reflection grating R having approximately 400 lines per mm. The inclina-

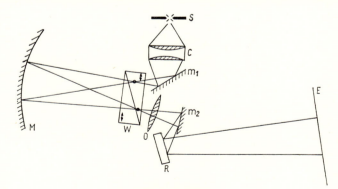

Figure 14.20

tions of the mirror and of the grating are so adjusted that the light falls almost normally on the grating. The grating reflects the light on to the observation screen E.

The Wollaston prism is at the centre of curvature of the spherical mirror and laterally displaced from the axis in such a way that the uniform colour observed on the screen is the white of higher order.

14.6.5   Observation of the convection currents produced by a candle (fig. 14.21)

The experimental set-up is that shown in figure 14.17. The Wollaston is placed at the centre of curvature of the mirror to obtain a uniform colour. The lighted candle is close to the mirror. The lens O forms an image of the candle on the screen E. By displacing the Wollaston along any of the two arrows, the image is seen in different colours.

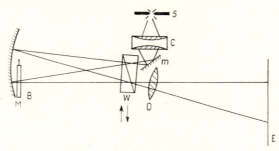

Figure 14.21

15

Study of Polarized Vibrations

The most general polarized vibration is elliptic. Its analysis consists of determining the ratio of the axes of the ellipse; their orientations and the sense in which the ellipse is described.

15.1 ANALYSIS OF A LINEAR VIBRATION

15.1.1 Principle

This involves determining its orientation or, more often, the variation in its orientation when it passes through an optical system. When the eye acts as the detector, the light beam is extinguished by rotating the analyser, the transmission direction of which is thus perpendicular to the direction of vibrations. This procedure is not precise. It is preferable to place, on one half of the light beam, an optical device which rotates the vibration OV to be studied, through an angle ε of a few degrees (fig. 15.1). When the transmission (OA) of the analyser is perpendicular to the bisector OB of the acute angle between the two vibrations, the two halves of the beam

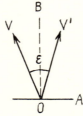

Figure 15.1

have equal intensities (weak) and the eye is sufficiently sensitive to appreciate this equality (half-shadow method).

A half-shadow plate may consist of a half-wave plate, placed on one half of a circular aperture (fig. 15.2) and the edge $x'x$ of which is parallel to one of the neutral lines, for example. A vibration V making an angle α with $x'x$ is transformed into a vibration V' which makes an angle $2\alpha = \varepsilon$ with V (cf. 13.6.1 C).

By this method one can determine the direction of the major axis of a highly eccentric elliptic vibration.

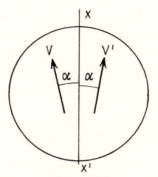

Figure 15.2

15.1.2   Setting up a half-shadow analyser (fig. 15.3)

a) Apparatus

  S: mercury vapour lamp, A_3 type, with a green filter.
  L: lens of about 30 cm focal length.
  P: polarizing sheet.
  D: circular diaphragm, 2 to 3 cm in diameter.
  C: half-wave plate for green light, placed on one half of *D* (fig. 15.2).
  A: polarizing sheet which can be fixed to *C*.
  L': lens with a focal length of 20 to 30 cm.
  E: white screen.

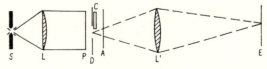

Figure 15.3

b) Adjustments

Distance between S and L: S is at the object focus of *L.*

Distance between D and L': L' forms an image of *D* on *E.*

Adjustment of P, D and A: P and *A* are crossed in the absence of *D.* On putting *D* in place, the light reappears in one half of the field covered by the half-wave plate. By rotating *D* in its own plane (*A* remaining fixed) it is shown that light can be extinguished in both halves of the field simultaneously; the neutral lines of the plate are thus parallel to *P* and *A.* The half-wave plate is then rotated through a small angle (5 to 10°) and is fixed to *A.* By rotating the ensemble, analyser plus half-wave plate, the two halves of the beam can be made equally intense. If a substance having rotatory power (cf. § 17.2) is introduced between the polarizer and the ensemble, analyser plus half-wave plate, the equality of intensity is disturbed. This can be restored by rotating the ensemble.

15.1.3   Setting up the experiment of analysis (fig. 15.4)

a) Apparatus

S: mercury vapour lamp, A_3 type, with a filter isolating the green line.

L: lens with a focal length of 20 cm.

D: circular diaphragm of 2 to 3 cm in diameter.

P and A: polarizing sheets.

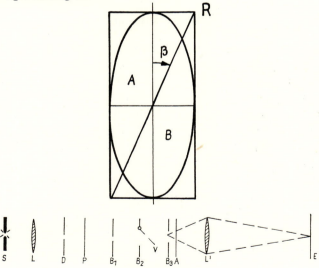

Figure 15.4

B_1: frame capable of supporting crystal plates and of measuring roughly the angles through which these are rotated.

B_2: frame with a flap V carrying a quarter-wave plate which can be rotated in its plane. The flap is pivoted on a hinge and can be moved out of the light path.

B_3: frame carrying a half-wave plate, which can be fixed to the analyser A.

L': lens with a focal length of 40 cm.

E: white screen.

b) Adjustments

Distance between S and L: such that the lens L gives an approximately parallel beam.

B_1 *and* B_2: B_1 and B_2 have no part in this experiment.

B_3: the circular diaphragm of 15.1.2 is placed in B_3.

Adjustment of L': L' forms an image of B_3 on the screen E.

15.2   ANALYSIS OF AN ELLIPTIC VIBRATION OF KNOWN ORIENTATION

15.2.1   Principle

Frequently the orientation of the vibration to be analysed is known a priori (cf. chap. 16). The determination of the ratio of the axes and the sense in which the ellipse is described is equivalent to the determination of the variations in the orientation of a linear vibration by the method of quarter-wave plate: if the vibration passes through a quarter-wave plate whose neutral lines coincide with the axes of the ellipse, it is transformed into a linear vibration R (fig. 15.4); the angle $\beta (< \pi/2)$ made by the linear vibration with the slow neutral line of quarter-wave plate is such that $\tan \beta = B/A$ (ratio of the axes of the ellipse) and it is described in the same sense as the ellipse (cf. 13.7).

One can, clearly, use the half-shadow method to determine the orientation of R. The experimental set up is that of 15.1.3 where B_1 and B_3 are empty to start with. B_2 carries a mica quarter-wave plate but the flap V is out of the light beam at the start. P and A are crossed. V is put in the light path and is rotated till extinction is restored; the neutral lines of the quarter-

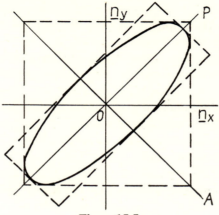

Figure 15.5

wave plate are thus parallel to P and A. V is now removed without rotating it. A mica plate ($\lambda/6$ to $\lambda/10$ for green) is placed in B_1. It is rotated in its plane till extinction is restored and is then given a further rotation of 45°. The axes of the ellipse coming out of the plate are thus parallel to P and A (fig. 15.5). It is shown that on rotating A extinction cannot be obtained, but only a minimum (when A is parallel to the minor axis of the ellipse). V is brought into the path of the rays (without rotating it): the neutral lines of the quarter-wave plate are parallel to the axes of the ellipse. Extinction can be obtained by rotating A, and this shows that the vibration coming out of the quarter-wave plate is linear. If the slow neutral line of the quarter-wave plate is known and also the indices n_x and n_y of the plate in B and its exact path-difference one can verify the rules concerning the sense in which the ellipse is described.

15.2.2 Setting up the experiment

See the set-up of 15.1.3.

15.3 ANALYSIS OF A VIBRATION IN THE GENERAL CASE

15.3.1 Principle

The method of quarter-wave plate is used

a) A simple analyser A is employed. By trial and error the orientations of the quarter-wave plate and the analyser are so adjusted that the neutral lines of the quarter-wave plate are parallel to the axes of the ellipse (this determines their orientation) and the vibration R is stopped by the analyser (this determines the ellipticity and the sense of the vibration).

b) A half-shadow analyser, consisting of A and a half-wave plate placed in B_3, is used. One proceeds as in a (Chauvin's method).

c) An analyser with four sectors is used; its principle is as follows (Chaumont's method). A quarter-wave plate MLN is superposed on a half-wave plate $M'L'N'$. Each of these plates possesses a straight border in the direction of a neutral line and covers half of the field. These two borders are perpendicular to each other (fig. 15.6). In the quadrants thus formed, the retardations are: 0, $\lambda/4$, $\lambda/2$, $3\lambda/4$.

This double plate is placed in B_3, in front of the analysing prism of fig. 15.4. Equal intensity in all the four quadrants can be obtained simul-

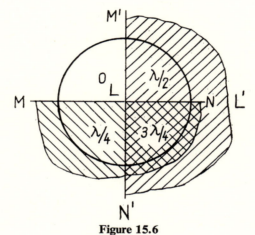

Figure 15.6

taneously only when the vibration coming out of the quarter-wave plate placed in B_2 is linear and is oriented along one of the neutral lines of the double-plate, MN for example. The frames B_1, B_2 and B_3 being empty to start with, extinction is obtained, then the double plate is placed at B_3 and it is rotated such that MN or $M'N'$ makes small angle (half shadow angle) with the direction of transmission of the analyser A. The linear vibration coming out of the polarizer P is found out as in 15.1.1. The quarter-wave plate is placed in B_2 and it is rotated till equality of the quadrants is restored. A crystal plate l is placed in B_1. It is found that the equality of the quadrants can be obtained by first rotating the ensemble $B_2 + B_3 + A$ (these elements are fixed together) till the quadrants $\lambda/4$ and $3\lambda/4$ have the same intensity (in this way one measures the angle α which defines the orientation of the axes of the ellipse coming out of l); and then, keeping B_2 fixed in the preceeding position, the ensemble $B_3 + A$ is rotated till the quadrants O and $\lambda/2$ are of equal intensity (this second operation furnishes the angle β of figure 15.4). Equality of the four quadrants is obtained by trial around the preceeding positions and precise values of α and β are thus determined.

15.3.2 Setting up the experiment

See the set up in 15.1.3.

16

Artificial Birefringence

16.1 PRINCIPLE

An isotropic medium subject to a vectorial or a tensorial action in a fixed direction, acquires anisotropy which manifests itself by birefringence. The medium acquires the properties of a uniaxial crystal, positive or negative in different cases, the optic axis of which has the direction of the applied action.

16.2 UNIFORM COMPRESSION OF GLASS

16.2.1 Principle

For all solid bodies, the difference in principal indices $\left| n_e - n_0 \right|$ produced in a parallelopiped subject to uniform pressure P, the light propagating normal to the direction of the applied effort, is given by:

$$\left| n_o - n_e \right| = CP\lambda \tag{16.1}$$

C is a constant. For crown, $n_0 - n_e$ is positive and C is of the order of 5×10^{-7} M.K.S. for yellow light.

A uniform pull produces a birefringence of opposite sign following the same rule.

16.2.2 Setting up the experiment (fig. 16.1)

a) *Apparatus*

S: carbon arc.

L: lens of any focal length.

P and A: polarizers, polarizing sheets. The field of view of the set up should be large and therefore polarizing sheets of about 20 cm in diameter should be employed.

B: isotropic system subject to actions which render it birefringent. In the present case it is a glass cube of 1 cm side (the glass cube is verified to be

175

isotropic before the experiment). Uniform compression can be applied by means of a pressing machine working with a screw; the machine is represented in section in figure 16.2. The metallic piece *M* slides parallel to the sides

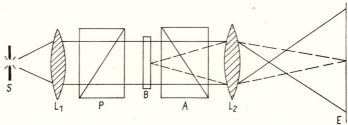

Figure 16.1

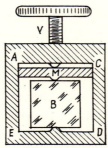

Figure 16.2

AE and *CD* of the rack. It serves as an intermediary between the screw *V* and the sample *B*, the pressed faces of which should be polished and made parallel to each other. By interposing sheets of card-board between lower face of *M* and upper face of *B* and between the lower face of *B* and the face *DE* of the rack, approximately uniform pressure is applied to *B*.

L_2: lens with a focal length of 40 cm.

E: white screen.

b) *Adjustments*

Distance between S and L_1: the lens L_1 gives a parallel beam of light.

Adjustment of B, L_2 and E: the lens L_2 forms an image of *B* on *E*.

Adjustment of P, A and B: the birefringence is sufficiently large in all cases to restore the light which was formerly extinguished by crossing *P* and *A* and in the absence of any stress on *B*. The directions of transmission of *P* and *A* should be oriented at 45° to the neutral directions of *B*.

16.3 PULLING OF A RUBBER SHEET

16.3.1 Principle

A sheet of rubber, sufficiently thin to be translucent, becomes strongly birefringent when pulled with the hands. In this case, the clustered molecules tend to unroll themselves in the direction of the pulling force.

16.3.2 Setting up the experiment

The set up is that of figure 16.1. The pressing machine and the glass cube are replaced by a sheet of rubber.

16.4 NON UNIFORM COMPRESSIONS AND TENSIONS

16.4.1 Principle

Transparent objects can easily be made from sheets of organic glasses (plexiglass) of 1 cm thickness. These objects can be subject to different stresses. If these stresses are applied normal to the plastic sheet through which light passes normally, one observes between crossed polarizers, dark lines (isoclinal lines) which are the loci of points where the neutral lines of the sheet are parallel to the transmission directions of the analyser and the polarizer.

16.4.2 Setting up the experiment

The set up is that of figure 16.1. One may study the flexion in a rectangular plate or the deformation of a hook on which a mass of 5 to 10 kg is suspended (fig. 16.3).

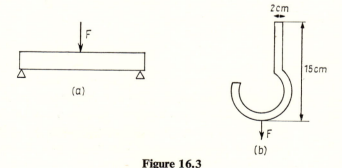

Figure 16.3

16.5 TEMPERED GLASS

16.5.1 Principle

The process of tempering is the rapid cooling of molten glass, the exterior parts of which become rigid and produce tensions and compressions in the mass which acquires a permanent birefringence. The plate glasses are very little tempered. The Securit glass is strongly tempered. It suffices to place a plate in the set up of figure 16.1, to observe bright colours at E. The glass tubes are tempered; to observe their birefringence these are immersed in benzene contained in a tank (not tempered) with parallel faces so as to reduce the effects of reflection and refraction.

16.5.2 Setting up the experiment

The set up is that of figure 16.1.

16.6 FLOW OF LIQUIDS

16.6.1 Principle

The existence of a uniform velocity gradient in a liquid consisting of stretched molecules may produce an orientation of the molecules and a birefringence.

The colloidal solutions of vanadium oxide $V_2O_5^\dagger$ are very suitable for the experiment.

Cellophane which is used as a wrapping material and as a support for adhesive ribbons, is obtained by laminating and coagulating the solutions of macromolecules. It shows a regular and quite large birefringence (the commercial adhesive ribbons are approximately half-wave for the yellow).

16.6.2 Setting up the experiment

The set up is that of figure 16.1. The birefringent system B consists of a tube T (fig. 16.4) of 1 cm cross-section connected, by means of rubber tubes, at its upper end to a funnel E, and at its lower end to an evacuating tube provided with a stop-cock. The tube and the funnel are filled with V_2O_5 solution and one waits till the liquid comes to rest. The flow which follows the opening of the stock-cock P restores the light. If the tube T is vertical, the neutral lines are at 45° to the vertical.

If the preceeding equipment is not available, the solution may be put in a parallel faced cell and it can be shown that on stirring the solution with a glass rod, light is restored between crossed polarizers. The isoclinal lines have irregular form.

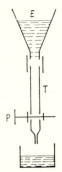

Figure 16.4

16.7 MAGNETIC BIREFRINGENCE

16.7.1 Principle

A) In general, crystals are anisotropic magnetically whether they are dia-or paramagnetic. In a uniform magnetic induction field B, approximately iso-diametric crystal fragments are acted upon by a couple which tends to bring the direction of maximum magnetic susceptibility parallel to the field B for paramagnetic crystals.

The first case is that of siderite $FeCO_3$, the principal susceptibilities of which have values: parallel to the axis $+140$ (C.G.S units); normal to the axis $+80$.

The second case is that of calcite: -4.1 parallel to the axis, -3.6 normal to the axis.

The phenomenon can be demonstrated by suspending in the air-gap of an electromagnet, a crystal connected to a cocoon string with a drop of resin, the ternary axis of the crystal being approximately horizontal. The string is left to untwist itself, in the absence of field, and one notes the period of oscillation under the influence of the very weak torsion couple of the string. On starting the current in the electro-magnet, one sees that the crystal takes on a new position of equilibrium (ternary axis parallel to field for siderite, normal to the field for calcite), and at the same time the period of oscillation about this position decreases very much, manifesting the existence of a strong couple.

B) Most of the molecules (with the exception of those which are mono-atomic or which belong to the cubic system) are magnetically anisotropic. Consequently, they tend to orientate themselves in a magnetic field, as the crystals in the experiment 16.7.1 (A), if the fluidity of the medium permits. The magnetic anisotropy is accompanied by optical anisotropy; the medium does not have the same refractive index for the vibrations parallel to the field and the vibrations perpendicular to the field; in this respect it behaves as an uniaxial crystal. The alignment of the molecules is disturbed by thermal agitation, therefore the birefringence decreases when the temperature increases. The difference in the principal indices is given by the formula:

$$|n_e - n_o| = C'\lambda H^2 \qquad (16.2)$$

where $C'(\lambda, T)$ is the Cotton–Mouton constant, λ is the wavelength and H is the applied magnetic field. The magnetic birefringence of pure liquids is too weak to be demonstrated by applying fields upto 1 tesla except by the use of extremely delicate methods of vibrational analysis and that too for individual observation only. The phenomenon can be observed in colloïdal solutions containing particles whose dimensions are much larger than the molecules and which are, therefore, less affected by thermal agitation; such a solution is placed between crossed polarizers and there is no light on the screen. When the liquid is subject to magnetic field, the light appears.

The sensitivity of the experiment is increased by using the half shadow method instead of the method of extinction. A double-plate is made from two identical cleaved sheets of mica (about $\lambda/10$) the neutral lines of which are crossed (cf. § 13.5). The polarizer and the analyser are crossed (fig. 16.1); the double plate is introduced in B, it is rotated till extinction is achieved and then it is further rotated through 45°. The lens forms an image of B on the screen E. A weak birefringence is added to one half of the double plate and is substracted from the other, with the result that the intensity in the two halves is different.

16.7.2   Setting up the experiment A

a) Apparatus

    S: carbon arc.

    L and L': two lenses with a focal length of 20 to 30 cm.

    E: white screen.

    crystals used: one cleaves with a knife two rhombohedrons (calcite and siderite), as regular as possible and with sides of about 0.5 cm. With a drop of resin (fig. 16.5) a silk thread of about 0.5 mm diameter, is fixed to one of the non-ternary summits (one of the six summits where all the three angles between the faces are not obtuse). The other extremity is fixed to a non-metallic support which enables the crystal to be placed in the air-gap of an electro-magnet.

electro-magnet: the field in the air-gap should be at least 0.5 tesla.

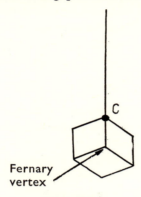

Fernary
vertex

Figure 16.5

b) Adjustments

Adjustment of S and L: the lens *L* gives an approximately parallel beam which illuminates the crystal.

Adjustment of L' and C: the lens *L'* forms a magnified image of the crystal on the screen *E*.

The lenses *L* and *L'* may be omitted and, instead, a silhouette of the crystal be made on the screen, *S* being many meters away.

16.7.3 Setting up the experiment *B* (fig. 16.6)

a) Apparatus

S: carbon arc.

L_1: lens with a focal length of 20 to 30 cm.

electromagnet: the field should be 0.8 to 1 tesla

C: cell made of non-magnetic material, about 1 cm long and of as small a width as possible (0.5 cm for example) so that the air-gap of the electro-magnet can be narrow and the field as intense as possible. Figure 16.7 shows

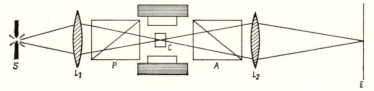

Figure 16.6

a model of an ebonite cell which can be easily made with a hack-saw and a file. The parallel faces are closed by cementing object slides with paraffin. The cell is filled with a colloidal solution of iron hydroxide.

P and A: polarizer and analyser, prisms or polarizing sheets.

L_2: lens of 20 to 30 cm focal length.

E: white screen.

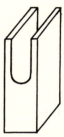

Figure 16.7

b) Adjustments

Adjustment of S, L_1 and C: the lens L_1 forms an image of *S* in the *C*.

Adjustment of P and A: the polarizers are crossed and their transmission directions are at 45° to the lines of the magnetic field.

Adjustment of C, L_2 and E: the lens L_2 forms an image of *C* on the screen *E*.

16.8   ELECTRIC BIREFRINGENCE

16.8.1   Principle

A) The dielectric crystals are, in general, electrically anisotropic. In a uniform electric field they are acted upon by a couple which tends to orientate them in such a way that the direction of their maximum permittivity be parallel to the electric field. For calcite, the relative principal permitivities have the values: parallel to the axis 8.0; normal to the axis 8.5. For rhombic sulphur the three principal permittivities are 3.7, -3.9 and 4.7.

B) The molecules are, in general, anisotropic and the induced dipole moment *P* which they acquire in an electric field *E* tends to align them parallel to the field. Furthermore some molecules possess a permanent electric dipole moment P_0 which also has the tendency to be orientated by the field *E*. These orientations, even though opposed by thermal agitation, suffice to render the fluid optically anisotropic, which results in a birefringence for the light travelling normal to the direction of the field E. The difference in the principal indices is given by the formula:

$$|n_e - n_o| = C\lambda E^2 \qquad (16.3)$$

where $C(\lambda, T)$ is Kerr's constant and λ is the wavelength.

16.8.2 Setting up the experiment A. Orientation of crystals in an electric field (fig. 16.8)

a) *Apparatus*

S: carbon arc.

L: lens of 20 to 30 cm focal length.

K: calcite crystal prepared as in the figure 16.5 or crystal of sulphur suspended in such a way that the direction of elongation of the octahedron be normal to the suspension threod.

CC': plane condenser of Aepinius with one plate earthed, and a voltage V applied to the other.

M: electrostatic machine (Van de Graaf or Felici).

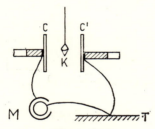

Figure 16.8

b) *Adjustments*

The voltage V and the distance between the plates are such that the field E should be at least equal to 3000 V/cm. The source and the lens are used as in § 16.7.2.

16.8.3 Setting up the experiment B. Kerr effect (fig. 16.9)

a) *Apparatus*

S: carbon arc.

C: ordinary condenser.

F: large slit (1 to 3 mm).

13*

L_1 and L_2: lenses of 20 to 30 cm focal length.

P and A: polarizer and analyser (prism or polarizing sheets) not annealed.

K: glass (not annealed) cell with parallel faces. 1 cm wide, 4 to 6 cm long, containing two electrodes (plates of copper) parallel to the large faces of the cell. The distance between the electrodes is 4 to 5 mm.

E: white screen.

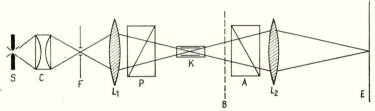

Figure 16.9

b) *Adjustments*

Adjustment of S, C and F: C forms an image of S on F.

Adjustment of L_1 and F: L_1 forms an image of F in K. K is filled with nitrobenzene and the electrodes are connected to the A.C. mains, 110 or 220 V (fig. 16.10) through a potentiometer ($R \cong 100\,\Omega$ for 2.5 A) and a

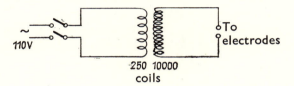

Figure 16.10

transformer T capable of giving a maximum voltage of 10,000 V (take necessary precautions for insulation).

Adjustment of L_2, K and E: the lens L_2 images the rear edges of the electrodes on the screen E.

Adjustment of P and A: P and A are crossed and their transmission directions are at 45° to the lines of the electric field. The voltage is progressively increased by changing R. The light reappears on the screen E.

By introducing a Bravais double plate at B, it can be shown that the phenomenon observed is certainly due to birefringence; at the same time the sensitivity of the experiment is increased. The lens L_2 projects the line of separation of B on the screen E.

17

Rotatory Polarization

A number of crystals, pure liquids and solutions exhibit the following properties: when a parallel beam of linearly polarized monochromatic light traverses these media the direction of vibration of the incident light is rotated through an angle α in the plane of the wave. The angle α is proportional to the thickness traversed, e. For an observer receiving the light, the rotation may be clockwise (dextro-rotatory substances) or anti-clockwise (laevo-rotatory substances).

This phenomenon can be demonstrated using the following set-up (fig. 17.1). A parallel beam of light issuing from a source S placed at the focus of a

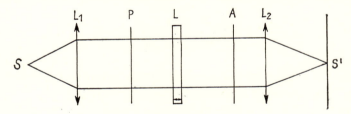

Figure 17.1

lens L_1, traverses a polarizer and a crossed analyzer. There is no light in the focal plane of L_2. If a plane parallel plate of a substance possessing the rotatory power, is placed at L, normal to the light beam, light is restored and it can be again extinguished by rotating the analyser through an angle α.

   The rotation of the plate in its own plane, or the interchange of the two faces of the plate do not bring about any modification. The experiments can be conducted with cubes of sodium chlorate, quartz plates cut perpendicular

185

to the optic axis (in birefringent crystals the rotatory power can be observed only in the direction of the optic axis; in other directions, it is masked by birefringence).

One may also employ tanks or tubes, with parallel faces, containing liquids. The angle α varies with the wavelength λ. For quartz and other transparent substances

$$\alpha_\lambda = Ae/\lambda^2 \quad (A = \text{specific constant of the substance})$$

If S is a source of white light, each of the monochromatic vibrations is rotated through α_λ. The analyser lets pass the components of these vibrations in its transmission direction. These components have different amplitudes and the sum of their intensities gives coloured light. If the analyser is rotated, the colours change. The colours corresponding to two perpendicular azimuths of the analyser are complementary.

17.2 ROTATORY POWER OF A QUARTZ PLATE CUT PERPENDICULAR TO THE AXIS

17.2.1 Principle

A quartz plate cut perpendicular to the axis is interposed in the set up of figure 17.1; the experiment is conducted with monochromatic or white light. Reappearance of the light, and the existence of colours are observed. It is verified that a calcite plate cut perpendicular to the optic axis does not restore the light.

17.2.2 Setting up the experiment (fig. 17.2)

a) *Apparatus*

S: carbon arc with a filter for observation in monochromatic light.

C: ordinary condenser.

L: quartz plates, cut perpendicular to the axis, of thicknesses 0.1 mm to 5 mm. There exists right-handed and left-handed quartz ($\alpha = 21°$ per mm for sodium light).

P and A: polarizer and analyser, two polarizing sheets.

O: lens with a focal length of 30 cm.

E: white screen.

b) *Adjustments*

Distance between S and C: this distance is such that the image of S is formed at about 40 cm from C.

Position of the plate L: the plate L, between the polarizers P and A, is at 5 to 6 cm from the condenser. To eliminate the phenomenon of birefringence, the light should be incident normally. In practice, the light beam is slightly convergent but this arrangement works well if the angle θ is not too big.

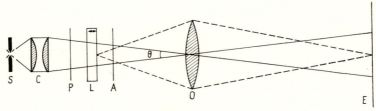

Figure 17.2

Adjustment of O: the lens O is at the image of S; thus maximum of light is obtained on the screen E. It forms an image of the plate on the screen E. 3 to 4 metres away.

c) *Remarks*

To show the complementary hues, the analyser A may be replaced by a calcite-glass prism. The plate L should be limited by a circular diaphragm an image of which is formed on the screen by the lens. With the same experimental arrangement, the rotary power of sodium chlorate, $NaClO_3$, crystals can be observed. These crystals are cubic and their rotatory power manifests itself in all directions. The rotation for yellow light is 3° per mm.

One can use, equally well, cells with parallel faces, 10 to 50 cm in length, filled with essence of turpentine ($\alpha \cong 0.3$ per mm) or a 100% solution of sucrose (ordinary sugar) for which $\alpha \cong 0.5°$ per mm.

17.3   ROTATORY POWER OF A SUGAR SOLUTION

17.3.1   Principle

A parallel beam of linearly polarized light illuminates a concentrated solution of sugar contained in a vertical tube (fig. 17.3) which contains a slight precipitate of Ag_2CO_3 (§ 9.10.2.2). The light diffused at right angles is observed. It is known that the angle of rotation is a function of the thickness traversed. The light vibration at a point z of the tube is in the yOx plane. The observer receives the ray Oy and, therefore, selects the regions of the tube where the vibration is parallel to Ox (cf. § 9.10.2).

In monochromatic light, bright and dark regions are observed along the length of the tube. When the polarizer is rotated, the maxima and the minima get displaced along the length of the tube, remaining equidistant. In white light, colours are observed. By rotating the polarizer, a coloured spiral appears which also rotates.

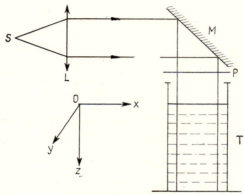

Figure 17.3

17.3.2   Setting up the experiment (fig. 17.4)

a) Apparatus

    S: carbon arc with filter for observation in monochromatic light.
    C: ordinary condenser.
    M: plane mirror
    P: polarizer (polarizing sheet).
    Solution A: concentrated solution of sugar in water (1 gm/cm^3).
    T: vertical glass tube, height 1 m, containing the solution.
    O: eye.

b) Adjustments

    Distance between S and C: the source *S* is very near to the object focus of *C*.

    Adjustment of M: the distance between *C* and *M* is arbitrary. The inclination of the mirror is so adjusted that light is incident normally on the sugar solution contained in the vertical tube.

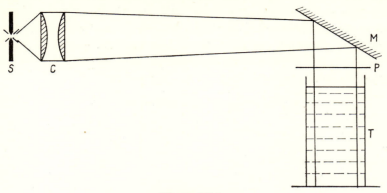

Figure 17.4

17.4   QUARTZ PLATE CUT PERPENDICULAR TO THE AXIS, TRAVERSED TWICE

17.4.1   Principle

We have seen (cf. § 17.1) that a quartz plate cut normal to the axis, rotates the direction of vibration of a linearly polarized light travelling along its axis. If by means of a mirror, the light beam is made to traverse the quartz plate a second time and in the reverse direction (fig. 17.5), it can be shown by placing the system between crossed polarizers that the extinction persists. The light vibration was rotated through an angle α on traversing the plate

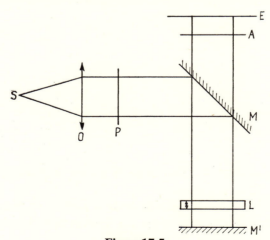

Figure 17.5

for the first time. On second passage the vibration has undergone another rotation through an angle α. But since the direction of propagation inside the plate was reversed, the rotation has also been affected in the opposite direction. Thus the beam falling on the analyser has the same state of polarization as the beam coming out of the polarizer.

17.4.2   Setting up the experiment (fig. 17.6)

a) Apparatus

   S: carbon arc.

   C: ordinary condenser:

   D: circular diaphragm.

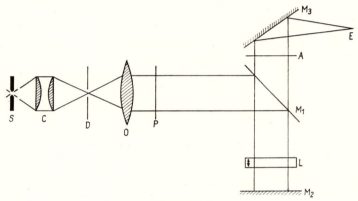

Figure 17.6

   O: lens of 30 cm focal length.

   P and A: polarizer and analyser; two polarizing sheets.

   M_1: semi-reflecting plate.

   M_2 *and* M_3: two plane mirrors.

   L: quartz plate of arbitrary thickness, cut normal to the optical axis.

   E: white screen.

b) Adjustments

   Distance between S and C: C forms an image of S on the circular diaphragm.

   Distance between D and O: the lens O forms an image of D on the screen E placed about 3 m away, the light rays having suffered reflections at the three mirrors.

Adjustment of M_1: M_1 is inclined at an angle of about 45° to the axis of the condenser so that the light beam is incident normally on the plate L.

Distance between M_1 and L: about 50 cm.

Distance between L and M_2: this distance is arbitrary; it is chosen for convenience.

Distance between M_2 and M_3: about 70 cm. The mirror M_3 reflects the light transmitted by M_1 on to the screen E.

Adjustment of P: the vibration incident on M_1 is polarized in the plane of incidence so that it does not get rotated on transmission.

17.5   CIRCULAR BIREFRINGENCE

17.5.1   Principle

Along any direction in an optically active medium, two circular vibrations of opposite senses can propagate independently, without deformation and with different velocities. On entering into such a medium a vibration of natural light is divided into two circular vibrations of equal amplitude which can be separated due to the difference in their refractive indices $|n_g - n_d|$.

The angle between the rays after they have traversed a prism in the minimum deviation position, is always very small. For a quartz prism of 60° (fig. 17.7) it is only of the order of 30″. To render the separation visible, a Fresnel's prism is employed. It consists of a number of quartz prisms, alternatively right-handed and left-handed, cemented with Canada balsam (fig. 17.8).

Let A be one half the angle of a prism. The angular separation is:

$$\varepsilon = N|n_g - n_d| \tan A \qquad (17.2)$$

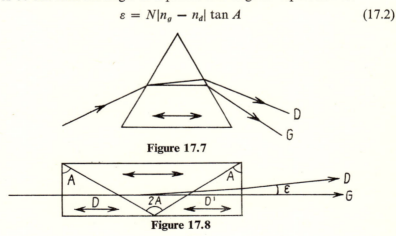

Figure 17.7

Figure 17.8

where N is the number of prisms constituting the Fresnel's prism. This prism is illuminated by a source at infinity. It is shown that it gives two images which are circularly polarized in opposite senses.

17.5.2   Setting up the experiment (fig. 17.9)

a) Apparatus

S: carbon arc.
C: ordinary condenser.
D: diaphragm with a very small circular aperture.
P: Fresnel's prism.
A: circular analyser.
L: lens with a focal length of 30 cm.
E: white screen.

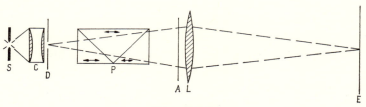

Figure 17.9

b) Adjustments

Distance between S and C: the condenser forms an image of the source at a distance of about 30 cm.
Distance between C and D: the diaphragm is placed next to the condenser.
Position of P: P is on the image of S.
Adjustment of D, L and E: the lens L forms an image of the circular aperture on the screen E, 3 to 4 m away.

17.6   CHANNELED SPECTRUM OF A QUARTZ PLATE CUT PERPENDICULAR TO THE AXIS

17.6.1   Principle

A source of white light at infinity illuminates normally a rod of quartz, the optic axis being parallel to the axis of the rod, placed between two polarizers making an angle θ between them (fig. 17.1). If the quartz plate is quite thick, the radiations composing the white light rotate through an angle α_λ

and all those for which the condition $\alpha_\lambda - \theta = (2K + 1)\,\pi/2$ is satisfied, are stopped by the analyser (fig. 17.10). The composition of the remaining radiations gives white light, which if analysed by a spectroscope gives a channelled spectrum.

If the quartz rod is rotated about its axis, the phenomenon does not change.

If the analyser is rotated about the emerging beam, the bands are displaced continuously. The sense in which the analyser has to be rotated to obtain a displacement of the bands towards the violet, determines the sign of the quartz.

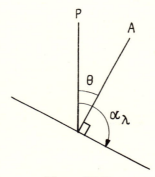

Figure 17.10

17.6.2   Setting up the experiment (fig. 17.11)

The set up is that of figure 14.10. It suffices to replace the quartz plate cut parallel to the optical axis by a quartz rod cut perpendicular to the optical axis, and of 2 to 3 cm thickness.

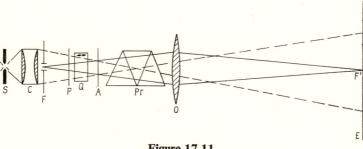

Figure 17.11

17.7 MAGNETIC ROTATORY POLARIZATION

17.7.1 Principle

Transparent isotropic substances placed in a uniform field of magnetic induction show rotatory polarization for the light beams propagating parallel
to the field. The rotation is proportional to the thickness e and the field B:

$$\alpha = \varrho Be \quad (\varrho = \text{specific constant of the substance}) \qquad (17.3)$$

The sign of α is related to the direction of the field. Contrary to the case
of natural rotation (cf. § 17.4), magnetic rotation is doubled by a double
passage through the medium in reverse directions.

The rotation is observed with the media which have a high value for the
constant ϱ: carbon bisulphide $\varrho = 0.042\ 10^6$ per Wb/m; dense flint
$\varrho \cong 0.07\ 10^6$.

If sufficiently intense and extended fields are available, the rotation can
be shown by the restoration of light between crossed polarizers (cf. § 17.1)
or by the extinction on rotating the analyser. Otherwise a biquartz, which
gives the tint of passage (sensitive tint), should be used between parallel
polarizers. The establishment of the magnetic field produces a difference of
hue between the two regions; on reversing the direction of the magnetic
field, the colours in the two regions interchange.

17.7.2 Setting up the experiment (fig. 17.12)

a) Apparatus

  S: carbon arc.

  C: ordinary condenser

  P and A: polarizer and analyser; two polarizing sheets.

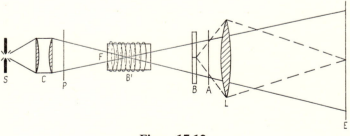

Figure 17.12

B: biquartz giving the tint of passage. This contsists of two plates of equal thickness, cut perpendicular to the optic axis, one right-handed and the other left-handed. A linear vibration incident normally is rotated by each plate through the same angle but in opposite senses. Generally the plates are made 3.75 mm in thickness, which produces a rotation of 90° for the mean yellow light. If the biquartz is placed between parallel polarizers, the yellow is stopped and both the plates show the same tint of passage. If the incident vibration undergoes a rotation, one of the hues becomes more blue and the other more red. In this way a rotation of the order of 1/5 of a degree can be detected.

L: lens of 15 cm focal length.

B': the removable coil of the transformer with 250 coils. It gives a magnetic field parallel to the light beam. Direct current (D.C) is given to the coil through a rheostat of maximum resistance equal to 40 ohms. There is a change-over switch in the electric circuit (fig. 17.13) to reverse the direction of the current and, consequently the direction of the field.

F: parallelopiped of very dense flint, cross-section 2 cm² and 5 to 6 cm in length. It is placed inside the coil such that the light traverses it along its length.

E: white screen.

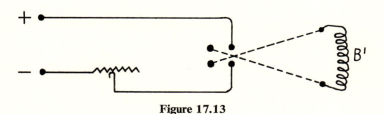

Figure 17.13

b) *Adjustments*

Distance between S and C: this distance is such that the condenser forms an image of the source at about 20 cm, at the centre of *F*.

Adjustment of B and L: the biquartz is placed between *E* and the lens. The lens *L* forms an image of the biquartz on the screen *E*, 3 to 4 m away.

Adjustment of B': the rheostat is worked very rapidly so that for a short duration an intense current flows through the coil.

17.7.3   Variation

The coil B' may be replaced by a small electromagnet; a hole is drilled along the length of its polar pieces (fig. 17.14). A current of 6 to 8 A may, for example, is passed through the coils of 500 turns for 15 to 30 seconds without danger. The rotation produced by the flint F can, thus, be seen without the biquartz.

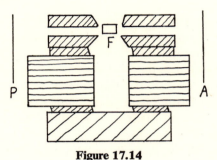

Figure 17.14

18

Dichroism

18.1 LINEAR DICHROISM

18.1.1 Principle

The difference in the refractive indices $|n - n'|$ of a birefringent substance for two linearly polarized monochromatic waves propagating in a given direction is related to the difference in the absorption coefficients $|K - K'|$ for the two waves.

Consider a plane parallel plate of dichroic uniaxial crystal with the optical axis parallel to the surface. A linear polarizer is placed before or after this plate. When the polarizer is rotated in its own plane, the light intensity transmitted by the combination varies; it passes through two extreme values when the transmission direction of the polarizer is parallel and perpendicular to the optical axis.

In white light, the dichroism manifests itself either by a difference in the intensity of the coloured light passing through the plate, or by a difference in its coloration.

18.1.2 Setting up the experiment (fig. 18.1)

a) Apparatus

$S:$ carbon arc.

$L_1:$ ordinary condenser.

$P:$ polarizer.

$C:$ crystal plate. We will use very thin ($e < 0.1$ mm) plates of green or pink tourmaline (rhombohedral) or thicker plates (1 mm or more) of trigonal crystal of ruby (Al_2O_3 coloured with chromium) or the quadratic crystals of the compound $CuCl_4(NH_4)_2$, $2 H_2O^\dagger$. The dichroism of green tourmaline is sufficiently large that a plate about 1 mm thick will transmit vibrations parallel to the axis only.

Birefringent prism or L: it is advantageous to observe simultaneously the absorption for the two vibrations parallel to the neutral lines of the plate by replacing the polarizer *P* with a birefringent prism, or by adding in one half of the beam and close to *C*, a half-wave plate *L*, whose neutral lines are at 45° to those of *C*.

L₂: lens with a focal length of 30 cm.

E: white screen.

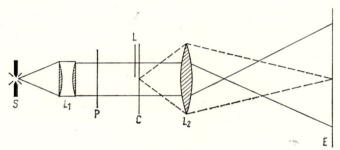

Figure 18.1

b) Adjustments

Distance between S and L₁: the condenser L_1 gives a parallel beam.

Adjustment of C, L₂ and E: L₂ forms an image of *C* on the screen *E*.

Position of P: P is rotated to observe the dichroism. The same observations can be repeated by placing *P* between *C* and L_2.

Adjustment of L: if a half-wave plate is used, its neutral lines are at 45° to those of *C*.

18.2   CIRCULAR DICHROISM

18.2.1   Principle

The difference in the refractive indices $|n_g - n_d|$ of a medium for right-handed and left-handed circularly polarized light, which is the cause of the rotatory power (cf. § 17.5), is related to the difference in extinction coefficients $|K_g - K_d|$ in certain absorption bands of the optically active substance.

Let Φ_0 be the flux of a parallel beam of monochromatic light, of wavelength λ, incident on a substance showing circular dichroism for this wavelength. Whether the circularly polarized light is right-handed or left-handed, the flux after traversing a thickness l of the substance is given by:

$$\Phi_d = \Phi_o \exp(-k_d l) \quad \text{and} \quad \Phi_g = \Phi_o \exp(-k_g l)$$

Φ_d and Φ_g are thus unequal.

18.2.2 Setting up the experiment for individual observation (fig. 18.2)

a) *Apparatus*

S: commercial sodium lamp.
L: lens of 10 to 15 cm focal length.

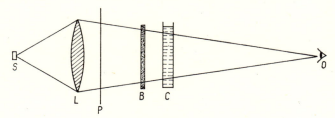

Figure 18.2

P: polarizer (nicol or polarizing sheet).
B: quarter-wave double plate. The eye is very sensitive to the difference between Φ_d and Φ_g when these fluxes illuminate two adjacent surfaces. One obtains this result by passing linearly polarized light through a double plate which is made as follows: a quarter-wave plate for the wavelength considered is made from a mica sheet. A straight edge parallel to one of the neutral lines is made and then the plate is cut into two, perpendicular to the edge; one of the two halves is rotated through 90° and the two pieces are juxtaposed. The neutral lines of the two halves are thus crossed and a linearly polarized light at 45° to the neutral lines will become right-handed circularly polarized on emerging from one of the two halves and left-handed circularly polarized from the other half.

C: container, 1 to 2 cm thick, containing a dichroic liquid[†]. For the liquid studied, and for sodium yellow light, the values are $K_g = 4.86$ cm^{-1} and $K_d = 5.12$ cm^{-1}.

O: eye.

14*

b) Adjustments

Adjustment of S, L and O: the lens *L* forms an image of *S* in the plane *O* where the eye is placed.

Adjustment of B and C: the eye is accommodated on *B* which is uniformly and brilliantly illuminated in the absence of *C*. In the presence of *C*, the two regions present a difference in intensity, always small but distinct. The solution in the container is diluted by trial and error till a good contrast is obtained.

Adjustment of P: the polarizer *P* is rotated through 90° with respect to the position at which its transmission direction is at 45° to the neutral lines. The illuminations in the two regions interchange.

18.2.3 Setting up the experiment for observation by projection
(fig. 18.3)

In this arrangement the circular dichroism manifests itself not only by a difference of intensity in the two halves of the field, but also by a difference in colour.

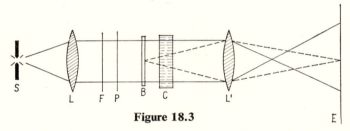

Figure 18.3

a) Apparatus

S: positive crater of a carbon arc.

L: lens of arbitrary focal length.

F: yellow filter or concentrated solution of potassium chromate to eliminate blue and green light for which the dichroism is weak or zero.

P, B and C: see § 18.2.2.

L': lens of 30 cm focal length.

E: white screen.

b) Adjustments

Adjustment of S and L: L furnishes an approximately parallel beam.

Adjustment of P, B and C: see § 18.2.2.

Adjustment of B, L' and E: the lens *L'* forms an image of *B* on the screen *E*.

18.3   SYNTHETIC LINEAR DICHROISM

18.3.1   Principle

Polarizing sheets are fabricated either by aligning coloured, strongly anisotropic molecules included in colourless crystals, or by dyeing with iodine
long-chain molecules of plastic which have been aligned by lamination
(cf. § 16.6).

The effect of alignment can be shown by the following experiment: a drop
of concentrated solution of methylene blue in alcohol is evaporated on
a glass plate. With the help of a polarizer the blue spot is shown (cf. § 18.2.1)
to be isotropic. It is then lightly rubbed, about ten times, with a piece of
cotton and always in the same direction. The spot becomes dichroic; it is
darker when the vibration is parallel to the direction of rubbing.

18.3.2   Setting up the experiment

The experimental set-up is that of figure 18.1 where the dichroic plate is
at C and L is suppressed.

19

Properties of
Electromagnetic Radiations

19.1   VISIBLE RADIATIONS

19.1.1   Emission spectra of metallic atoms

The spectral composition of a mercury vapour lamp and of the flame of an arc containing vapours of potassium is studied. The light coming from the source is split by one of the methods indicated in chapter 1. Different radiations constituting the incident light are observed on the screen (fig. 19.1).

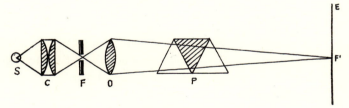

Figure 19.1

a) Apparatus

S: mercury vapour lamp of medium pressure or a source with potassium vapours obtained by boring two vertical carbon rods of an arc and filling them with the powder of a potassium salt (KBr or KCl) or with pieces of pure potassium.

C: ordinary condenser of 12 cm diameter.

F: vertical slit of adjustable dimensions.

O: lens of 35 cm focal length and of a diameter large enough to collect all the light issuing from *S*.

P: direct vision prism.

E: white screen.

b) Adjustments

Distance between S and C: *C* forms an image of *S* at a distance of about 40 cm from itself.

Distance between C and F: *C* forms an image of *S* on *F*.

Distance between F and O: about 20 cm. The image *F'* of *F* given by *O* should be a few meters away from *O*. If *O* has a focal length of 30 cm, *FO* should be greater than this value.

Distance between O and P: about 15 cm. The prism *P* is placed so that the whole of the light beam can pass through it.

Distance between P and E: 4 to 5 meters. *O* forms an image *F'* of *F* on the screen *E*. When a carbon arc is used, the height of the slit is adjusted so as to suppress the continuous spectrum of the carbon rods.

19.1.2   Absorbtion spectra

19.1.2.1   Principle

The flux *F* of monochromatic radiation transported by a quasi-parallel beam decreases on traversing a thickness *x* of a substance, according to the law:

$$F = F_0 \exp(-Kx)$$

F_0 being the incident flux, for $x = 0$; K is the coefficient of absorption, characterizing the substance (for a given frequency and a given temperature); it has the dimension inverse of length. The transmission coefficient is the ratio of the emerging flux to the incident flux. The absorption of most substances is strongly selective: it varies very much with frequency. We will observe selective absorption spectra in the visible, ultraviolet and the infra-red.

19.1.2.2   Setting up the experiment (fig. 19.2)

a) Apparatus

    S: tungsten ribbon lamp.

    L_1, L_2: lenses of 10 to 20 cm focal length.

    F: slit.

    P: prism (see no. 1.3).

E: white screen.

C: tank with parallel faces containing the absorbing media. It is convenient to give it the form shown in figure 19.3, separating it into two parts with a glass diagonal *G*. One of the two halves is filled with the absorbing solution, and the other with the solvent so as to eliminate the deviation of the rays produced by the absorbing prism.

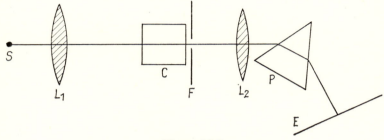

Figure 19.2

b) Adjustments

The ribbon *S* is horizontal. L_1 forms an image of *S* on *F* (horizontal) which should cover the entire width of the entrance face of the tank *C* of figure 19.3. L_2 gives an image of *F* on *E* which becomes a spectrum when *P* (with its edge horizontal) is interposed. The effect of an increasing thickness of the absorbing substance can, in this way, be observed along the height of the spectrum.

Figure 19.3

c) Interesting spectra

Aqueous solution of copper sulphate (red is absorbed); of cobalt salts (green is absorbed); of nickel salts (red and blue are absorbed); of neodymium salts (five bands in yellow, green and blue); of potassium permanganate (band in the green which is resolved into five bands very close together); oxyhaemoglobin obtained by putting a few drops of blood in one cm³ of distilled water (two bands in the green yellow and in red); alcoholic solution of chlorophyll, obtained by maceration of spinach leaves dried at 120°C (band in red).

Gaseous nitrogen dioxide: a few chips of copper are placed at the bottom of a tank (with parallel faces) and a little fuming nitric acid is poured over them. The tank is covered with a glass plate. Crimson coloured vapours are released which give a spectrum containing very many bands in the blue and the green.

The colour obtained by transmission when a film of an absorbing substance is illuminated in white light, may depend on the thickness traversed. A substance whose coefficient of spectral transmission, for a certain thickness, is represented by curve I of the figure 19.4, for example, transmits according to curve II under a 4 times larger thickness; the dominant colour is thus displaced towards red. The solution of indigo carmine[†] in water or that of bromophenol blue[†] in water, slightly alkalised with sodium bicarbonate, are blue in thin films and violet red in thick films.

A mixture of substances absorbing complementary regions of the visible spectrum, is opaque. A solution of cupric chloride in diluted hydrochloric acid is poured in a flask with parallel walls and a solution of methyl red[†] in amyl alcohol is then added. The first solution transmits only green and blue, the second only red and orange. The solutions are not miscible. If the flask is shaken to emulsify the two solutions, the liquid appears black.

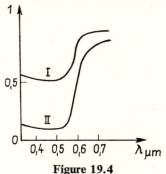

Figure 19.4

19.2 ULTRAVIOLET RADIATIONS

19.2.1 Ultraviolet spectrum

The radiations existing beyond the violet do not give the sensation of colour to the eye and are absorbed by glass: these are ultraviolet radiations. Their existence can be verified by using a quartz prism (transparent for these wavelengths) and receiving the spectrum on a fluorescent screen. The ultraviolet radiations extend from about 4000 to 100 Å.

19.2.2 Setting up the experiment (fig. 19.5a)

a) Apparatus

S: electric arc or mercury lamp of medium pressure with a quartz envelope, or an iron arc.

F: vertical slit.

M: concave mirror of 20 cm diameter and of 60 cm radius of curvature.

P and P': two identical quartz prisms with refracting angle of 60°.

E: the observation screen *E* is a white screen on which fluorescent paper
has been fixed; one straight edge of the paper is horizontal and is placed
at mid-height of the spectrum so that one can compare the extensions
of the ultraviolet and visible spectra of the source employed. It is advantageous to use a quartz lens for projecting on *F* an image of *S*, and to replace
the mirror *M* by a quartz lens.

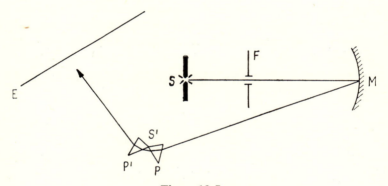

Figure 19.5a

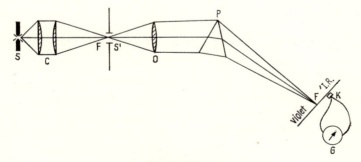

Figure 19.5b

b) Adjustments

Distance between S and F: the slit is very near the source *S* but sufficiently far away that is does not get overheated.

Distance between F and M: the mirror is placed at 35 cm from *F*.

Distance between M and P: the prism is placed at *S'*, the image of *S* given by the mirror *M*.

Adjustment of the prisms: the prisms are set for minimum deviation with respect to the beam coming from *M*. The prism *P'*, placed just after *P*, increases the dispersion of the spectrum given by *P*.

Distance between P' and E: the screen *E* is placed at *F'*, the image of *F* given by the mirror *M*.

19.2.3   Absorption of ultraviolet radiations

The set up is that of figure 19.5.a. Plane parallel plates of various substances are interposed at *F*: ordinary glass (opaque below 0.35μ), quartz, fused quartz, calcium fluoride (transparent up to 0.2μ), Wood's glass containing nickel oxide (opaque in the visible and transparent between 0.39μ and 0.35μ).

19.3   INFRARED RADIATIONS

19.3.1   Infra-red spectrum

Just as there are radiations beyond violet to which the eye is not sensitive, there are radiations beyond red which are invisible to the eye: these are the infra-red radiations, above 0.8 microns roughly. To verify their existence, the spectrum of an electric arc is explored with a thermocouple connected to a galvanometer. In the visible spectrum, the emission is weak and the galvanometer is not deflected. When the thermocouple is positioned beyond the red region, in the near infra-red, the radiation increases and the galvanometer is deflected.

19.3.2   Setting up the experiment (fig. 19.5)

a) Apparatus

   S: electric arc.

   C: ordinary condenser.

   F: vertical slit of adjustable dimensions.

   O: lens of 33 cm focal length.

   P: glass prism with a refracting angle of 60°.

   R: detector; this is a thermocouple *K* connected to a galvanometer.

b) Adjustments

Distance between S and C: the image S' of S formed by C is at 30 cm from C.

Adjustment of F: the slit F is placed at S'. Its height is such that the entire spectrum is received by the thermocouple.

Distance between F and O: F is at about 30 cm from O. The image F' of F as given by O should be quite far away and should be quite luminous.

Adjustment of P: the prism P, adjusted for minimum deviation position, is at about 20 cm from O.

Distance between P and K: the image F' of F is at about 70 cm from P. The spectrum F' is explored by displacing the thermocouple K which is carried on a travelling support.

19.3.3   Absorption of infra-red radiations

Instead of using the spectrometric arrangement of figure 19.5b, it is easier to use the bulk of non-dispersed infrared radiations emitted by a carbon arc or by an incandescent lamp of type I3. The detector may be a thermocouple, or a Crookes radiometer the blades of which turn when heated by radiations, or a thermometer based on the expansion of gas, consisting of a thin-walled ball filled with air and fitted with a tube containing a liquid marker, I, (fig. 19.6).

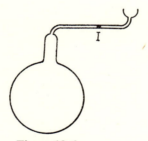

Figure 19.6

One of the above detectors receives the radiation from the source and registers its response. If a tank of water with parallel faces, is interposed the response is considerably reduced; water possesses, in fact, numerous intense absorption bands throughout the infra-red region. If the tank is filled with a solution of iodine in carbon tetrachloride dark violet colour, and which absorbs almost all the visible radiation but transmits the infra-red, the response remains strong.

19.4   ULTRA-HERTZIAN RADIATIONS
19.4.1   General

The analogy between light waves and hertzian waves can be demonstrated by using wavelengths between one centimetre and one decimetre. The following experiments of photometry, geometrical optics and physical optics are conducted with ultra-hertzian waves. A source of hertzian waves differs from an ordinary source of light in the sense that it emits a monochromatic coherent and linearly polarized vibration. Many commercial devices are available for demonstration experiments. They differ in regard to the frequency of the emitted waves and the mode of production. The smaller waves have a more precise directivity; they need relatively smaller instruments and the accessories are less cumbersome, but they are more expensive. The following types are available, for example:

	Frequency MHz	Wavelength cm	Production	Transmitter	
I	2,450	12	Triode	antenna	Cenco (USA)
II	5,850	5	Magnetron	cornet	Leybold (German)
III	8,500–9,500	3.1 to 3.5	Klystron	cornet	Phillips
IV	30,000	0.9	reflex		Phillips

The receiver is a half-wave antenna (I) or a horn (III-IV), the detector a semi-conductor crystal (Si, Ge) diode. The current is measured by a micro-ammeter; it may, after amplification, work a loud speaker with the help of which one can hear the sound of the mains frequency.

The instructions for the use and adjustment of the transmitter and the receiver are given by the manufacturers.

19.4.2   Radiation diagram of the emitter

Principle

This diagram is analogous to the indicator diagram for the emission of a light source. For a half-wave antenna, the radiation diagram is a figure of revolution about the axis $Z'Z$ of the antenna. The modulus of the electric field of the emitted wave in a plane passing through $Z'Z$ and at a distance large compared to the wavelength, varies as shown in figure 19.7. For a cornet, the radiation diagram in a horizontal plane resembles the right-hand loop of the figure 19.7; Ox being the axis of the cornet.

Setting up the experiment

The receiving antenna is displaced on a circle of about 1 m radius, at the centre of which lies the emitting antenna. The two antennae are, to start with, normal to the plane of the circle, the indication of the receiver remains constant in all the positions. When the antennae are in the plane of the mirror, the response of the receiver passes through a maximum along x or x' and is zero along z or z'.

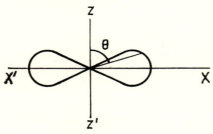

Figure 19.7

19.4.3   Some experiments of geometrical optics

Principle

The laws of reflection, of refraction and the formulae of Fresnel (chap. 10) are applicable to hertzian waves. For each frequency, the dielectrics are characterised by a real index of refraction.

Setting up the experiments

A)   The transparency of dielectrics (cardboard, glass, wood, paraffin, ebonite) and the opacity of conductors (metals, solutions of electrolytes, animal tissues) is verified by interposing plates of these substances between the transmitter and the receiver suitably aligned.

B)   For hertzian waves the reflection coefficients of metals are very close to unity. The transmitter producing a quasi-parallel beam, the laws of reflection are roughly verified when turning the mirror M (a leaf of aluminium or of copper 1 m²). The response of the receiver is maximum when it lies on the geometrical path of the reflected beam (fig. 19.8).

If the antenna of the transmitter is placed at the focus of a metallic parabolic mirror (fig. 19.9) a quasi-parallel beam is obtained.

C)   With a paraffin prism whose section is a right-angled triangle and whose faces are at least 40×40 cm², one can verify the refraction of a quasi-parallel beam (fig. 9.10a) and its total internal reflection for an incidence of 45° on the hypothenuse face (fig. 19.10b).

In the process of total internal reflection, the flux of radiation energy penetrates in the less refringent medium up to a depth of about one wave-

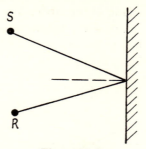

Figure 19.8

length, before returning to the original medium. Total internal reflection can be suppressed by bringing a second paraffin prism in contact with the first, along their hypotenuses. The hertzian beam traverses, of course, the plane parallel plate so formed; when the two faces are moved away from each other, keeping them parallel (fig. 19.11), the receiver gives a response which becomes weaker and finally imperceptible when the distance between the prisms is equal to a few wavelengths.

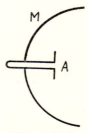

Figure 19.9

D) For centimetric waves lenses are made with paraffin. The diameter is about 30 cm, the focal length is determined by the radii of curvature of the faces. A maximum response is obtained when the receiver is placed a the geometrical image of the transmitter (fig. 19.12). When a cornet transmitter is placed at the object focus of such a lens, the directivity of the beam is very much improved.

E)   Reflection on a plasma. The index of refraction of an ionized gas is less than unity, decreasing with the frequency of the waves. The phenomena of total internal reflection are easily obtained with centimetric waves.

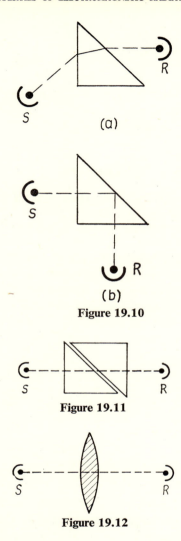

(a)

(b)

Figure 19.10

Figure 19.11

Figure 19.12

The plasma consists of a gas which fills a neon tube or a fluorescent tube of illumination. The disposition of the transmitter and the receiver with respect to the tube T (vertical and maintained by a wooden support) is as shown in figure 19.13; the distances ST and TR are of the order of 0.5 to 1 m. The response of R is very weak when the tube is turned off. It increases considerably when the tube is lighted.

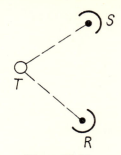

Figure 19.13

19.4.4　Diffraction by a slit

Principle: see chapter 9.

Experimental set up

The slit is a rectangular aperture, 2 to 3 wave lengths wide and 4 to 5 wavelengths high, cut in a metal foil of area 0.5 to 1 m². The axis of the transmitter S is oriented along the normal to the centre of the slit F (fig. 19.14). By displacing R on a circle C centered on the slit, variations of intensity analogous to those of figure 9.2, are observed.

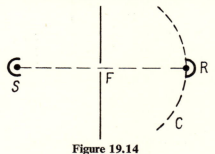

Figure 19.14

Adjustment

S gives a quasi-parallel beam. The distances SF and FR are of the order of 10 wavelengths.

19.4.5　Diffraction by circular apertures – Fresnel's zones

Principle

If a point transmitter S and a point receiver R (fig. 19.15) are aligned and separated by a distance D, the various points of the spherical wavefront Ω of radius r and centre S send vibrations which interfere at R. The spheres of centre R and radii $RP = r'$, $r' + \dfrac{\lambda}{2}$, ..., $r' + K\dfrac{\lambda}{2}$, divide the wavefront Ω into Fresnel's zones whose radii are given by:

$$\varrho_k = \sqrt{\frac{K\lambda rr'}{D}}. \tag{19.1}$$

The vibrations received at R from two successive zones are in phase opposition and are of almost equal amplitude.

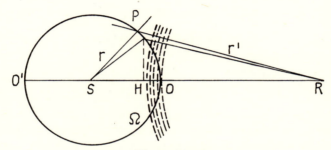

Figure 19.15

Experimental set up

The formula (19.1) shows that for $r = r' = D/2$, if D is of the order of 1 m, λ a few centimetres, the value of ϱ is a few centimetres for the first zone. A disc of radius ϱ_1, and 3 annular rings of outer and inner radii (ϱ_2, ϱ_1), (ϱ_3, ϱ_2) (ϱ_4, ϱ_3) are cut from a metal foil (fig. 19.16). These pieces are fixed concentrically by means of nails on a plywood support which is then suspended vertically.

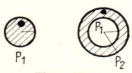

Figure 19.16

Adjustments

S gives a divergent beam. S and P are aligned at the calculated distance. The support of the zones is held vertically and at equal distances from S and P, the centre of the zones being on the straight line SP. All the zones, being in place, R does not give any response. On removing the central disc, a signal is obtained. On removing the second zone also, the signal falls to

15*

almost zero (the distance RP may be slightly adjusted, if necessary, to obtain this result). The removal of the third zone increases the signal, and it diminishes again when the fourth zone is removed too.

19.4.6   Interference between sources of very small dimensions

A) Young's fringes

Principle: see § 2.2.

Experimental set up

A system of two parallel slits is constructed as in 19.4.4; these are $\dfrac{\lambda}{2}$ in width and 2λ apart, for example. Three interference maxima on either side of the central maximum can be observed.

Adjustments: see figure 19.14.

B) Different sources

Principle

On account of the coherence of hertzian sources, the preceeding experiment can be carried out by using two different, but synchronous, transmitters.

Experimental set-up

An antenna transmitter may supply power, with the help of flexible coaxial cables, to two identical and parallel antennae placed at a distance of the order of 4 to 5 wavelengths; the two antennae will be in phase.

Adjustments

The two antennae S_1 and S_2 (fig. 19.17) give quasi-parallel beams which are oriented to obtain a common region. The detector is placed at a distance, D, of about 20 wavelengths from the middle point of the straight line joining the two transmitters. It is displaced along $x'x$. The distance between two adjacent maxima or two adjacent minima of the response is the fringe spacing. The wavelength λ can be measured in this way

$$\lambda = i\,\frac{d}{D}$$

C) Lloyd's mirror

Principle

A point source S radiates obliquely on to a plane mirror MM' (fig. 19.18). At a point such as R, a direct ray interferes with a reflected ray which appears to be coming from a point S', symmetrical to S with respect to the mirror. The central fringe is at R, on the prolongation of the mirror. It corresponds

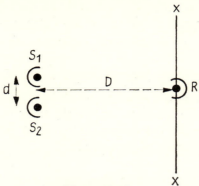

Figure 19.17

to destructive interference proving, thereby, that the reflection has introduced a phase change of Π

Experimental set-up

The mirror M is a metal plate of dimensions large compared to the wavelength.

Adjustments

The distance RM of the detector from the mirror is of the order of 10
to 20 wavelengths. The distances SH and SM are adjusted by trial so as
to obtain a suitable fringe system, which is detected by displacing R along RP.
A minimum is obtained at R.

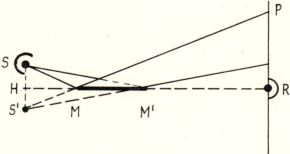

Figure 19.18

D) Stationary waves

Principle

Plane monochromatic electromagnetic waves, reflected normally from a
plain metallic mirror, interfere with the incident waves and give a system
of stationary waves, which are formed of nodal planes (electric field is zero

at any time) parallel to the mirror and separated by $\dfrac{\lambda}{2}$, and anti-nodal

planes (amplitude of the electric field is maximum) alternating with the preceeding ones.

Experimental set up

The transmitter S gives a quasi-parallel beam. The mirror M (fig. 19.19), a 1 m² metal plate, is placed normal to the beam at a distance of about 30 wavelengths from S. The detector R is, in all cases, a small antenna; L is a wooden scale graduated in millimeters.

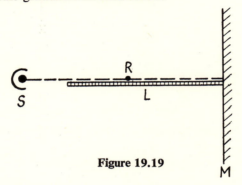

Figure 19.19

Adjustments

R is disposed parallel to the electric field emitted by S. It is displaced along L. The distance between two adjacent minima is equal to $\dfrac{\lambda}{2}$.

19.4.7   Michelson's interferometer

Principle: see chapter 4.

Experimental set up

The beam splitter G (fig. 19.20) should consist of a glass plate or a series of parallel insulated wires, separated by about $\dfrac{\lambda}{5}$ and stretched on a wooden frame. The mirrors M_1 and M_2 are metal plates. L is a graduated scale. G, M_1 and M_2 should measure at least 60×60 cm. All the supports should be wooden ones.

Adjustments

M is placed at about 1 m from S and normal to the axis of the beam emitted by S. M_2 is placed on the normal to SM_1 drawn at the point I; the distance $IM_2 = IM_1$ and M_2 is oriented normal to M_1. G is placed at I and is oriented at 45°. R is placed on the extension of the line $M_2 I$. The displacement of M_1 along L, without changing its orientation, produces variations in the response of R which have a period of $\dfrac{\lambda}{2}$.

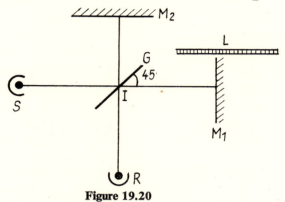

Figure 19.20

19.4.8 Multiple-beam interference. Fabry–Perot interferometer

Principle: see chapter 6.

Experimental set up

Glass has a high refractive index (2 to 2.5) for hertzian waves. Its coefficient of reflection is, therefore, large (0.3 to 0.4 under normal incidence). The set of two thin glass plates G_1 and G_2, (fig. 19.21) maintained parallel to each other, constitutes a multiple-beam interferometer.

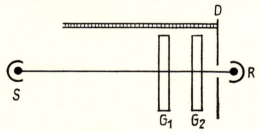

Figure 19.21

Adjustments

The transmitter S gives a quasi-parallel beam. The plates G_1 and G_2 are placed normal to the direction SR; D is a metallic diaphragm limiting the field to its central region only. The displacement of G_1 along the graduated scale L, varies the response of R with a period of $\dfrac{\lambda}{2}$.

19.4.9   Selective reflection of stratified media

When light penetrates into an inhomogeneous dielectric medium in which the coefficient of reflection varies periodically (with a suitable period), one observes selective reflection of certain radiations for which the reflected vibrations are in phase.

Experimental set up (fig. 19.22)

a) Apparatus

Source S: the source is a transmitter of centimeter waves ($\lambda = 3$ cm).

Reflecting surfaces: surfaces weakly reflecting centimeter waves are made by stretching sheets of paper on wooden frames (40×40 cm). Six mirrors of this type, M_1 to M_6, are vertically mounted on the joints of a system of jointed rhombi, made of wood or metal; the separation of the mirrors can easily be varied by this arrangement.

R: detector of centimeter waves.

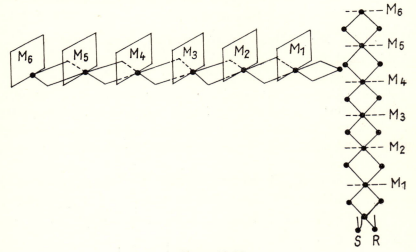

Figure 19.22

b) Adjustments

Adjustment of S and R: the transmitter S and the detector R are placed side by side, facing the mirrors.

19.4.10   Polarization of hertzian waves

A)   The emitted waves are linearly polarized. The electric field is parallel to the direction of the antenna or to the large diameter of the horn opening.

This is verified by rotating the antenna or the receiving horn around the axis along which they are aligned with the transmitter. When the transmitter and the receiver are parallel, the response of the latter is maximal. It is zero when they are perpendicular to each other.

B) *Experiment of Fresnel and Arago*

Synchronous electromagnetic vibrations do not interfere if they are perpendicular to each other. This is verified with the arrangement of the experiment 19.4.6.b. On turning one of the two emitting antennae so that it is normal to the other, the interference phenomenon disappears.

C) *Brewster's angle:* see § 11.3

The vibration being in the plane of incidence, no reflection takes place from a dielectric surface (large glass plate, hypotenuse face of the prism used in the experiment 19.4.3 c) when the angle of incidence is such that its tangent is equal to the refractive index n of the dielectric. It is difficult to get a null response of the detector (parallel to the transmitter). But if in the position of minimum response, the transmitter is rotated through 90° in its own plane so as to render the emitted vibration normal to the plane of incidence, the detector (rotated through 90° as well) will give a much stronger response (n is close to 1,5 for paraffin, and 2 to 2.5 for glasses).

D) *Artificial dichroism*

A set of parallel conducting wires separated by a distance of about a centimetre and stretched on a wooden frame of 50 × 50 cm constitutes a dichroic plate (chapter 18). The electric field of an electromagnetic wave is absorbed much more when it is parallel to the wires than when it is normal to the wires.

This is verified by introducing the frame with wires between the transmitter and the detector, properly aligned, and then rotating it in its own plane.

20

Thermal Radiation

20.1 RADIATION FROM BLACK AND GREY BODIES

20.1.1 Principle

The electromagnetic radiations emitted by bodies maintained at an absolute temperature T, excluding all other causes of excitation, obey certain simple laws. Black body, an ideal concept, is a body for which the coefficient of absorption A_λ is equal to 1 for all radiations. The energy of the radiation for a wavelength λ is designated by L_λ. L_λ is a function of λ and T only. It is given by Planck's formula:

$$L_\lambda = \frac{C_1}{\lambda^5} \; \frac{1}{\exp\left(\dfrac{C_2}{\lambda T}\right) - 1} \tag{20.1}$$

C_1 and C_2 are constants.

From this formula one can derive Stefan's law: the total energy of a black-body

$$\int_0^\infty L_\lambda d\lambda$$

is proportional to T^4; and Wien's law: the product $\lambda_m T$ is a constant where λ_m is the wavelength at which L_λ is maximum for a black body at temperature T. Those bodies for which A_λ is less than unity and is independent of λ, are known as grey bodies. The curve representing the spectral distribution of energy for these bodies is the same as for a black body, except for a change of scale.

The black body and numerous incandescent sources follow Lambert's law: their luminance L is the same in all directions.

20.1.2.1   Stefan's and Wien's laws

20.1.2.1 *Principle*

a) Tungsten radiates as a grey body. By raising its temperature, it is shown qualitatively:

1) that the spectrum as a whole becomes more brilliant (Stefan).

2) that it extends further into blue and violet (Wien).

b) The colour of an incandescent source depends on its temperature (Wien). The emitted light becomes more and more "white" as the temperature is raised.

1) carbon-filament lamp ($T = 2000°C$).

2) lamp with a tungsten filament in vacuum ($T = 2500°C$).

3) lamp with a tungsten filament in an inert gas ($T = 2700°C$).

20.1.2.2   Setting up the experiment (fig. 20.1)

a) *Apparatus*

S: incandescent lamp with a straight ribbon or filament (I_2 or I_1) the temperature of which can be varied by varying the heating current with a rheostat.

L: lens forming an image of the vertical ribbon *S* on the screen *E*.

P: prism with its edge vertical.

b) *Adjustments: see* § 1.3.

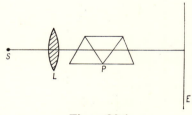

Figure 20.1

20.1.3   Lambert's law

20.1.3.1   Principle

A small area *S* surrounding a point *P* on the surface of a source (fig. 20.2), gives an intensity *I* in a direction *P* making an angle θ with the normal to *S*:

$$I = LS \cos \theta$$

the source is limited by a diaphragm D of small area Σ normal to the direction Px. If the source follows Lambert's law, the intensity, I, does not vary with the inclination of the source since its surface used is: $S = \dfrac{\Sigma}{\cos\theta}$

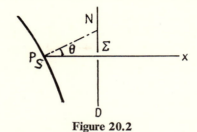

Figure 20.2

20.1.3.2 Setting up of the experiments

A) The area of a source is varied by using a tungsten ribbon lamp and rotating it about the axis of the ribbon. The radiation falls on a thermocouple connected to a galvanometer. The deflection is measured as a function of the angle through which the ribbon is rotated. It is proportional to $\cos\theta$ (fig. 20.3).

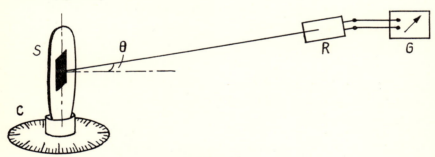

Figure 20.3

a) Apparatus

S: lamp with a vertical tungsten ribbon.
C: graduated circle capable of rotating about a vertical axis.
R: thermocouple.
G: projection galvanometer.

b) Adjustments

R is in a horizontal plane passing through the middle of S and at such a distance from S that the deflection of G be quite large when the ribbon is normal to SR. The deflection is, then, maximal. Starting from this position, C is rotated through 15°, 30°, 45°, ... and the respective indications of G are noted.

B) A source consisting of a black surface, 10 to 15 cm in diameter, is chosen. It is limited by a circular diaphragm of 3 to 4 cm in diameter. The radiation falls on a thermal detector connected to a galvanometer. When the source is inclined, the observed deviation remains constant (fig. 20.4).

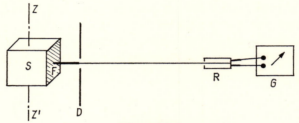

Figure 20.4

a) Apparatus

    S: Leslie's cube (see § 20.2.1.2)
    D: circular diaphragm.
    R: thermocouple.
    G: projection galvanometer.

b) Adjustments

    S is placed on a rotating table whose vertical axis of rotation passes through the middle of a blackened face of the cube. The centres of F, D and R are aligned.

20.2 KIRCHOFF'S LAW

Principle

    The spectral distribution of energy radiated by a body whose coefficient of absorption is A_λ, is given by Kirchhoff's law:

$$I_\lambda = A_\lambda L_\lambda \tag{20.2}$$

20.2.1 Coefficient of absorption A

20.2.1.1

Principle

    A blackened surface absorbs more than a polished surface, all other things being equal.

Experimental set up

Two plates of duraluminium (15 × 15 cm²), 1 cm thick are held vertical, parallel to each with a distance of 2 to 3 cm between them. This is achieved by introducing the lower edges of the plates in two slits made in a wooden brick. One of the two surfaces facing each other is polished (D_1), the other D_2 is covered with lamp black. Two wooden balls B_1 and B_2 are stuck at the middle of the exterior faces with grains of soft wax (fig. 20.5). A cylinder of pig iron or brass, suspended by an iron wire, is introduced between the plates quite symmetrically, without touching them. This cylinder had been brought to dark red heat in a Meker flame. After a few instants, the ball B_2 falls, the wax having melted. After a few minutes, it can be shown by a touch of hand that the external face of D_2 is very hot, and that of D_1 is only warm.

Figure 20.5

20.2.1.2.

Principle

The formula (20.2) shows that at a given temperature, the luminance B of a surface emitting temperature radiations is proportional to its coefficient of absorption, all other things being equal.

Setting up the experiment

1) *Leslie's cube:* the entire emission of equal surfaces, raised to the same temperature and disposed identically with respect to a thermal detector, is compared. The source is a cubic metal box, open on top, capable of being turned about a vertical quaternary axis $Z'Z$. One of the external faces is dull black; the opposite face is painted with white lead; the intermediate faces are polished.

The radiations fall on a thermal detector. The replacement of a blackened face by a polished face results in a considerable decrease of the deflection (by 9/10 for example). The white face radiates almost as much as the black face at wavelengths emitted principally at 100 °C ($\lambda_m = 7.7\ \mu$ according to Wien's law). The experimental arrangement is analogous to that of figure 20.4.

a) Apparatus

S: cubic box.

P: thermopile disposed on the normal to the blackened surface at its centre.

G: galvanometer.

b) Adjustments

S is filled with boiling water; the distance of *P* from *S* is so adjusted as to obtain a high deflection of *G*. Galvanometer deflections for the radiations from different faces are noted.

2) *Thermal emission of cobalt oxide:* in a pyrex test tube, a few crystals of cobalt nitrate, dehydrated in a drying oven at 105°, are introduced. By inclining the tube, these crystals are held on the wall close to the bottom. In this position the tube is heated in a Bunsen burner, till the salt decomposes emitting NO_2 and leaving a black adhesive spot of cobalt oxide on the wall. The tube, so prepared, is strongly heated in a Meker burner: the transparent glass does not emit visible radiations at a temperature at which the oxide is dark red.

Similarly, one can heat a piece of a refractory and absorbing crystal (tourmaline, black mica) in a quartz tube.

3) *Thermal emission of iodine vapour:* a few crystals of iodine are introduced in a quartz tube which is evacuated and sealed. On strongly heating in the colourless flame of a Meker burner, the tube fills with iodine vapour, dark violet in colour, and emits a red light visible in the dark. Under the same conditions, transparent gas does not emit any visible radiation (water vapour and carbon dioxide emit the same infra-red radiations as they absorb).

20.2.1.3

Principle

The equation (20.2) shows that a body absorbs the same thermal radiations as those it emits (if $l_\lambda \cong 0$, $A_\lambda \cong 0$, because $L_\lambda < l_\lambda$).

If the radiations of a black-body at T'^0 and with luminance L' traverse a body at $T^0(T' > T)$, the emerging radiations are composed of a transmitted component and an emitted component:

$$L_\lambda + L'_\lambda(1 - A_\lambda) = A_\lambda L_\lambda + L'_\lambda(1 - A_\lambda) = L'_\lambda - A_\lambda(L'_\lambda - L_\lambda)$$

Since $L'_\lambda > L_\lambda$, the preceeding expression is less than L'_λ. Thus the emerging radiations are weaker than those in the spectrum of the black body at T'^0.

Experimental set up

A) *Absorption by carbon of radiations emitted by a carbon arc*

A Bunsen burner B (fig. 20.6) is supplied by gas which has been passed through benzene in a gas washing bottle so as to charge it with its vapour. The ferrule of the burner being closed, the flame is very bright. If the shadow of this flame is projected on a screen, by using a small hole illuminated by a carbon arc as the source, this shadow will be dark. The light of the flame is due to the presence of incandescent carbon particles (by covering the flame with a cold metal surface, a deposit of lamp black is obtained). At 1500°C these particles absorb the brilliant light which is emitted by the arc at 3500°C.

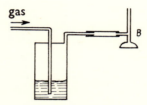

Figure 20.6

The blue flame of a Bunsen burner fed on oxygen (with ferrule open) does not give a shadow; it contains water vapour and carbon dioxide which are transparent to visible light.

B) *Absorption and emission of sodium lines: reversal of lines*

The spectrum of a continuous arc is formed on a screen and in the path of the rays dense sodium vapour is produced. Under these conditions, a broad dark line is observed in the spectrum at the position of the D lines (fig. 20.7).

If the arc is masked, the yellow line, weakly emitted by the vapour, appears at the same position on the screen.

a) Apparatus

$S:$ carbon arc.

$C:$ condenser forming an image of S on the hole T.

$L_1:$ lens forming an image of T at T' (fig. 20.7b, elevation) above the Meker burner M

$L_2:$ lens forming an image of T' on the slit F.

$L_3:$ lens forming an image of F on the screen E.

P_3: direct-vision prism.

K: iron small spoon, 5 to 6 mm in diameter, obtained by pressing a piece of metal sheet and fixed through an extension Q (fig. 20.7 c) to a support which holds it at the base of the flame of M, below T'.

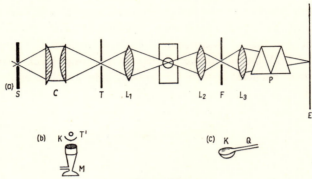

Figure 20.7

b) *Adjustments*

The burner is lit with a non-luminous flame, and a continuous spectrum is obtained. A piece of metallic sodium, the size of a pea, is placed in K; it melts and after a few moments, ignites, colouring the flame intensely.

The use of metallic sodium gives a high density vapour resulting in a broad absorption line and permits the use of a dispersive system P of mediocre resolution and also a wide slit F: the continuous spectrum is not quite pure but it is brilliant.

To render the phenomenon conveniently visible and to avoid, at the same time, the spreading of the sodium smoke produced by its combustion in the hall, M should be enclosed in a metal box drilled with holes for the light to pass.

c) *Absorption and emission of the Welsbach burner*

The mixture of cerium and thorium oxides, which forms the Welsbach burner emits almost as a black-body in the blue. If a non-lighted Welsbach burner is illuminated with an arc, a brilliant image excited by the blue and violet radiations from the arc, is observed on the screen. This image is greatly reduced when the burner is lighted, because it absorbs the blue radiations of the arc which it normally emits (fig. 20.8).

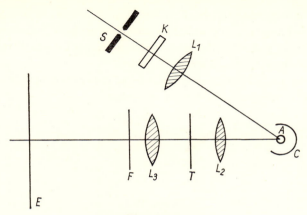

Figure 20.8

Apparatus

S: carbon arc.

K: cell containing an ammoniacal solution of copper sulphate.

L_1: lens forming an image of the crater of S on the fabric of the Welsbach burner A.

L_2: ($f = 10$ cm) forms an image of the illuminated region of A on the hole T.

L_3: ($f = 15$ cm) forms an image of T on the fluorescent screen E (barium plato-cyanide or quinine acid sulphate) across the blue glass F.

$C:$ a half-cylinder of metal used as a protection against the light of the Welsbach burner.

20.2.1.4   Emission of dichroic bodies

Principle

An anisotropic body which absorbs unequally linear vibrations of different orientations (cf. n° 18.1) may emit them through thermal radiation, with unequal intensity. The equality of the ratio L/A has been verified for the principal vibrations in tourmaline.

Setting up the experiment

A 1 mm thick plate of black mica, taken from a sample whose dichroism has been verified by transmission, is used; its lower face is heated in the

16*

blue flame of a Bunsen burner *B* (fig. 20.9) till the upper face becomes dark red. On examining this face with a birefringent prism *P*, in such a way that the two images given by the principal vibrations be juxtaposed, a clear difference in intensities is observed.

All samples will not be found suitable for this experiment which cannot be projected, neither directly nor by television.

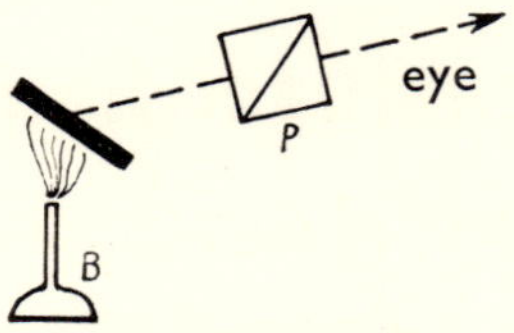

Figure 20.9

21

Absorption and Emission of Radiations by Atoms and Molecules

21.1 PHOTOELECTRIC EFFECT

21.1.1   Introduction

All substances irradiated by an electromagnetic radiation of sufficiently high frequency are ionized, which means that they lose some of their electrons. For atoms and molecules, there are a number of values for the ionization energy W_i corresponding to the successive loss of a number of electrons.

The frequency v_i required to produce the ionization is such that: $W_i = hv_i$. h designates Planck's constant. For metals, in which electrons are not strongly held, the frequency v_i is situated in the visible or the ultraviolet: the wavelength corresponding to this frequency is called the photoelectric threshold.

21.1.2   Photoelectric effect in zinc

21.1.2.1   Principle (fig. 21.1)

The photoelectric threshold for zinc lies in the ultraviolet. A zinc plate, Zn, is attached to the cap of an electroscope. An ultraviolet source, S, is placed at a short distance from the plate. The source is masked. If the electroscope is charged negatively with the help of a glass rod rubbed on silk, it is discharged immediately on uncovering the source. The source is again masked and the electroscope is charged positively with an ebonite rod rubbed on cat fur. The charge remains when the source is uncovered. Thus the metal loses negative charge under the influence of light. The demonstration may be completed with two electroscopes E_1 and E_2. One is charged positively and the other negatively (fig. 21.1 a). The radiation from S discharges both of them; the electrons lost by one are attracted by the other. In fact, of the experiment is repeated by placinga metallic screen between E_1 and E_2, only E_1 gets discharged (fig. 21.1 b).

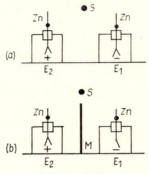

Figure 21.1

The existence of a photoelectric threshold may be demonstrated by interposing a glass plate between S and the negatively charged electroscope. Since glass absorbs ultraviolet, on uncovering the source, the electroscope will not be discharged. If the ordinary glass is replaced by Wood's glass, the discharge does take place.

21.1.2.2 Setting up the experiment (fig. 21.2)

a) Apparatus

S': carbon arc or a projection lamp.

L_1: lens of 30 cm focal length.

E: gold-leaf electroscope, with two opposite windows in the case enabling the projection; it carries a zinc plate, Zn, 5×5 cm² approximately, which has been freshly cleaned with emery paper.

S: Hg lamp with a quartz envelope.

L_2: lens of 40 cm focal length.

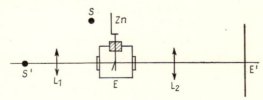

Figure 21.2

E': white screen.

l: plate of ordinary glass.

l': plate of Wood's glass.

b) *Adjustments*

The lens L_1 gives a parallel beam. The source S is placed at a short distance (9 to 10 cm) from the plate.

The lens L_2 projects the leaves of the electroscope on the screen E'. l or l' is interposed between S and the zinc plate.

21.1.3   Ionization by X-rays

An electroscope is charged and placed at a few metres from the source of X-rays. When the source starts emitting the X-rays, the charge is lost rapidly, whatever its sign: this is because the ionization of the molecules in air produces positive and negative ions which are free to move and are attracted by the charged electroscope.

21.2   MEASUREMENT OF PLANCK'S CONSTANT

Various spectral radiations, emitted by a mercury vapour lamp and isolated by filtering or by dispersion, are made to fall on a photoelectric cell of a particular type. Under the action of the radiation of frequency v_1, the electrons of mass m and of charge e emitted by the photo-cathode acquire a kinetic energy $y_2\, mv_1^2 = hv_1 - W_E$, which depends on the frequency v_1 of the incident radiation (W_E is the energy required to extract an electron). This kinetic energy is measured by a method of opposition: a retarding voltage, V_1, is applied between the photo-cathode and the collector of the electrons and it is so adjusted that the fastest electrons just cease to reach the collector. The following relation holds under these conditions:

$$eV_1 = 1/2\, mv_1^2 = hv_1 - W_E.$$

For an incident radiation of frequency v_2, the retarding voltage V_2 required to stop the fastest electron is such that:

$$eV_2 = hv_2 - W_E.$$

From the measurements of V_1 and V_2, h can be determined:

$$h = \frac{e(V_1 - V_2)}{v_1 - v_2} \qquad (21.1)$$

The value of e is 1.9×10^{-19} C. The frequencies of mercury lines employed in the experiment are ν_1(blue) $= 6.88 \cdot 10^{14}$ s^{-1}; ν_2(green) $= 5.49 \cdot 10^{14} s^{-1}$.

21.2.2 Setting up the experiment

a) Apparatus

The cessation of the photoelectric current can be precisely determined only by a very sensitive instrument, such as a mirror galvanometer associated with an electronic tube to increase the sensitivity. We have used the Leybold equipment (detailed instructions for use are supplied with each instrument: amplifier, micro-ammeter and photoelectric cell). The photo-sensitive film is a layer of potassium with a metallic coating on the external side. (A metal ring situated at a distance of about 1 cm from the cathode acts as the collector of electrons).

b) Details of the electrical apparatus (fig. 21.3)

    C: photo-cell.
    B: 4-volts battery to supply the retarding voltage.
    V: D.C. voltmeter, 0 to 2 volts, to measure the retarding voltage.

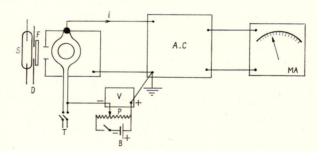

Figure 21.3

    P: potentiometer to vary the voltage *V*.
    MA: micro-ammeter.
    AC: D.C. amplifier.
    T: circuit to heat the anode to eliminate impurities.
    S: mercury vapour lamp A_3 type.
    D: iris diaphragm.
    F: interference filter.

c) Adjustments

1) to clean the anode, the circuit T is closed for a few seconds, the current being about 1 A.

2) D.C. amplifier is switched on.

3) A known monochromatic radiation is directed onto the photo-cell by placing at F a filter to isolate the green line and then the blue line, without changing the geometry of the light beam as defined by D. The micro-ammeter is deflected.

4) The retarding voltage is increased from 0 to V till the reading of the micro-ammeter is zero. Two different values of V are obtained for the two different radiations.

21.3   IONIZATION POTENTIAL

21.3.1   Ionization potential of mercury vapour

21.3.1.1   Principle

An electron with kinetic energy equal to, at least, $W_i = eV_i$ may on colliding with a vapour molecule M, remove an electron from it. The presence of the ions M^+ thus formed is revealed by their participation in the conduction current in the vapour. W_i has a definite value for a given type of molecule M.

Mercury vapour is contained at low pressure in a triode (fig. 21.4) whose heated filament emits electrons. They are accelerated by an adjustable po-

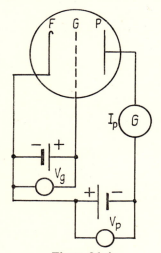

Figure 21.4

tential difference V_g applied to the grid G, and are repelled by the plate P which is at a negative potential V_p. On increasing V_g from zero upwards the galvanometer g does not indicate any current I_p in the plate circuit till V_g attains the value $V_i = 10.1$ volts (which is the ionization potential for the mono-atomic molecule Hg) (fig. 21.5), since the positive ions Hg^+ are then produced and are attracted by the plate.

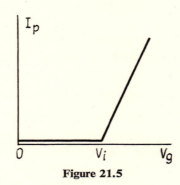

Figure 21.5

21.3.1.2   Setting up the experiment

a) *Apparatus*

T: thyratron, type RL 17 or RL 150.

V_g: grid tension, variable between 0 to 15 V, furnished by a battery of accumulators and a potentiometric arrangement, or a variable grid supply available commercialy.

V_p: constant plate voltage of the order 6 volts. It should be possible to reverse it.

The voltages V_g and V_p are read on voltmeters which have graduations of large dimensions, or are projected onto a screen.

G: D.C. ammeter, 20 to 50 mA, capable of being projected.

b) *Adjustments*

The filament is heated. To start with, the plate is given positive potential; V_g being zero, the normal electronic current of the triode is observed on G and its sense is noted.

The sense of V_p is reversed. G does not show any current. V_g is progressively increased; when it attains the value 10.1 volt, G shows a deflection in the opposite sense to that which was observed earlier.

21.4   EXPERIMENT OF FRANCK AND HERTZ

21.4.1   Principle

In a triode containing mercury vapour at low pressure, the heated filament F emits electrons (fig. 21.6). They are accelerated by a known and adjustable potential difference V_G, applied between F and the grid G, which gives them a kinetic energy $\frac{1}{2} mv^2 = eV_G$.

The plate P is given a potential V', slightly less than V_G, so that electrons with low velocities cannot reach the plate. When V_G is progressively increased from zero upwards, the plate current i_p, measured by a galvanometer g, increases and then decreases abruptly at $V_1 = 4.9$ volts, increases again up to $2V_1$, decreases, etc. ... (fig. 21.7). This phenomenon can be

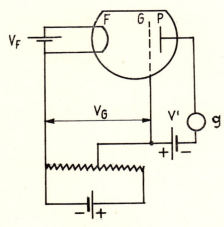

Figure 21.6

explained on the assumption that the collision between electrons and mercury atoms, for electron energies less than eV_1, are elastic. For the value eV_1, the energy is absorbed by the atoms which jump to an excited state, and then return to the fundamental state by emitting a radiation of frequency v such that $hv = eV_1$. The electrons which have lost their kinetic energy eV_1 in this way, can no longer pass through the interval GP. On increasing V_G, the electrons are again accelerated and may reach the plate with an energy eV_1. They can also undergo an inelastic collision.

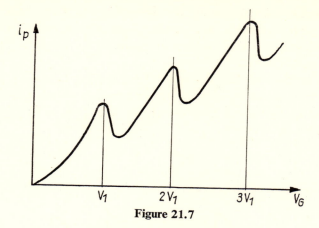

Figure 21.7

21.4.2   Setting up the experiment

a) Apparatus

For this experiment we use a mercury vapour tube (tetrode) produced by the firm Leybold (fig. 21.8). The voltages V_F and V_G are stabilized.

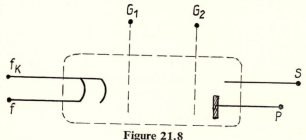

Figure 21.8

b) Adjustments

The required pressure of the mercury vapour is obtained by heating the tube in a furnace to about 200°C. The current i_p can be measured by a micro-ammeter associated with a D.C. amplifier (cf. 21.2.2). To make the experiment more impressive, the curve of figure 21.7 should be produced on an oscillograph screen. For this purpose, the voltage producing the horizontal scanning of the oscillograph (time base, fig. 21.9) is applied to the grid G_2 after a suitable attenuation by a potentiometer; V_{G_2} is of the order of 30 volts at the maximum. (The results are satisfactory with a frequency < 100 Hz.) If necessary, a condenser may be used to suppress the zero-frequency component.

The potential difference across a resistance in the plate circuit, is fed to a low frequency amplifier, the output of which is given to the terminals of the oscillograph. Thus a voltage proportional to the plate current i_p is fed to the oscillograph.

At each instant, the abscissa of the spot is proportional to V_{G_2} which starting from 0, varies linearly during one period, and the ordinate is proportional to i_p. By varying progressively the amplitude of V_{G_2} through the potentiometer Q, the curve $i_p = f(V_{G_2})$ can be drawn.

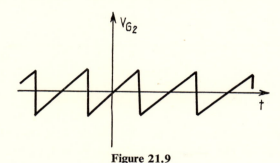

Figure 21.9

21.5 ZEEMAN EFFECT

21.5.1 Principle

When atoms are subjected to a uniform magnetic field, the spectral lines emitted or absorbed by them are generally modified: a single line is replaced by a number of lines, called the Zeeman components, whose frequencies differ very little from the original line; they can be resolved by a high resolution spectrometer. In the simplest cases, where there is a transition into two simple spectral lines, three lines (normal triplet) are observed in a direction normal to the lines of the field: the central line has the frequency v_0 of the original line; the two lateral lines lie on either side of the central line and are separated from it by Δv. These three lines are linearly polarized: the central line parallel to the direction of the magnetic induction B, the side components in an azimuth normal to the preceeding one (fig. 21.10b). In the longitudinal direction, i.e. in the direction of the field, only two components are observed; they are circularly polarized in opposite senses (fig. 21.10c). The separation Δv is proportional to B:

$$\Delta v = 1 \cdot 41 \times 10^6 \, B \text{ (Hertz)} \qquad (21.2)$$

For $B = 1$ Tesla, $\Delta\nu$ is of the order of 1 MHz. A high resolution spectrometer is, therefore, necessary to separate the Zeeman components in an easily obtainable field, B. A standard Fabry–Perot is quite suitable.

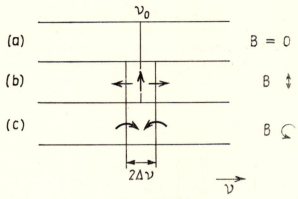

Figure 21.10

21.5.2   Setting up the experiment (fig. 21.11)

a) Apparatus

E: electro-magnet capable of producing a field of 0.5 to 1 Tesla in an air-gap of 2 to 1.5 cm. The old instruments of Ruhmkorff* are quite suitable. The air-gap should be adjustable and a canal should be drilled in the polar pieces if observation is to be made along the field. If the current is not more

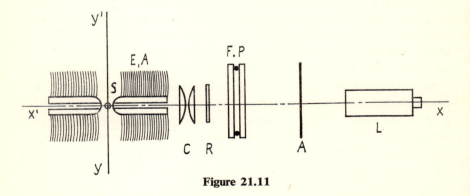

Figure 21.11

---

* Among modern instruments, the small model constructed by Beaudoin is sufficient.

than a few amperes, cooling by running water is not necessary during the limited duration of one observation.

S: cadmium vapour lamp of A_2 type. To reduce space, the glass envelope of circular section surrounding the discharge tube may be given an elliptic orm by heating it suitably; or it may be removed entirely (though it is detremental to the life of the lamp) thus permitting the reduction of the air-gap to 1.5 cm.

C: ordinary condenser.

R: red glass isolating the radiation $\lambda = 6438.7$ Å which is a single line.

F.P.: Fabry–Perot (§ 6.2) with half-silvered plates separated by a distance of 5 mm. Its resolving power is of the order of 5.10^5.

A: linear analyser (polarizing sheet) or a circular one (§ 13.7).

L: telescope of a goniometer set for infinity (or unaccommodated eye, or objective of a television camera adjusted for infinity).

b) Adjustments

Distance between S, C and F.P.: C forms an approximately equal sized image of *S* on the *F.P.* The *F.P.* is adjusted for parallelism with the naked eye to start with, and then with *L* through which one can observe the circular fringes of equal inclination centred on the optical axis. *R* and *A* are taken away during the adjustments. *R* being placed in front of *F.P.*, on increasing progressively the current in *E*, each ring enlarges progressively and then splits into two. In the scheme of the figure, the interposition of *A* in front of the telescope objective, does not vary the relative intensities of the two systems of rings when *A* is a linear polarizer; if *A* is a circular polarizer (polarizing sheet plus a quarter-wave plate oriented at 45°), one of the two systems produced due to splitting disappears; if the quarter-wave plate alone is turned through 90°, the other system disappears whereas the former one reappears. The adjustments and the procedure of observations are the same when *C, R, F.P., A* and *L* are aligned along *yy'* instead of *xx'*, but a single ring is split into three rings when the current is increased; the diameter of the central ring retains its initial value. Further, the linear analyser *A* stops the central system of rings when its transmission direction is parallel to lines of the field, and the two displaced systems of rings when it is rotated through 90°.

Observation

The resolving power required for observation along *xx'* is half that required for the direction *yy'*, the field being equal in the two cases.

Variation

J. L. Cojan (Caen) uses for the transversal observation an electromagnet of magnetron and a standard of 1 cm thickness.

21.6 PHOTO-CHEMICAL ACTION

21.6.1 Principle

Vision and photography which are the two most important detectors of radiation, are photo-chemical phenomena. The former is reversible: the visual pigment on the retina suffers a transformation under the action of light and regenerates its initial form in darkness. Such an action is imitated in the following experiment: the molecule of methylene blue in water solution is activated by visible light; it fixes hydrogen by becoming colourless at the same time the Fe^{++}ions are oxidised to Fe^{+++}. In darkness, the Fe^{+++}ions are reduced at the expense of the hydrogen of the discoloured compound, which regains its blue colour.

21.6.2 Experimental set-up (fig. 21.12)

a) Apparatus

S: carbon arc.
D: diaphragm.
L: f is 10 to 20 cm.
C: container with parallel faces for the solution, 1 to 2 cm thick.
E: white screen.

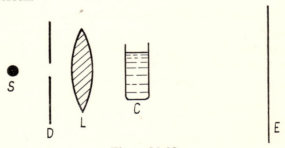

Figure 21.12

b) Adjustments

L forms an image of D on the screen E. The diameter of D is such that the entire light beam passes through the solution. In a few seconds, the image at E fades. S is masked for 10 seconds: the image becomes blue again and the process repeats itself.

21.7 PHOTO-CONDUCTIVITY

21.7.1 Principle

The conductivity of some solids increases when they are subjected to a radiation of frequency sufficient to transfer the electrons from the valency band to the conduction band. Zinc sulphide shows this effect under the action of visible light.

21.7.2 Experimental set up (fig. 21.13)

a) Apparatus

E: gold leaf electroscope (see 21.1.2).

M and M': plane metallic electrodes; one is connected to the leaves and the other to the case of the electroscope.

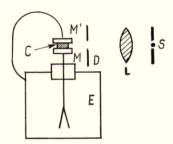

Figure 21.13

C: crystal of zinc sulphide (parallel faced fragments can easily be cleaved).

S: carbon arc.

L: converging lens.

D: circular diaphragm.

b) Adjustments

Distance between S and C: L forms an image of *S* on *C*. The diameter of *D* is such that the crystal alone receives the light. The crystal is placed in position and the electroscope is charged (positively or negatively). The charge is kept while the crystal is not illuminated and disappears rapidly on uncovering the source.

22

Luminescence

The term luminescence means the emission of radiation by some bodies; its cause is not simply a rise in their temperature as is true in the case of emission by incandescence. A more precise distinction between luminescence and incandescence is based on Kirchhoff's law (20.2): if the emission of a body is higher than that of a black body at the same temperature, it emits through luminescence. A prefix specifies the kind of process by which the luminescence is produced.

22.1 ELECTRO-LUMINESCENCE

22.1.1 Principle

An electric field of sufficient intensity suitably applied to a body may excite it to emit light. In all cases, the colour of the light emitted can be produced by incandescence only at a temperature considerably higher than that of the emitter.

22.1.2 Experimental set-up

In Geissler tubes, a gas or a vapour at a pressure of the order of 0.01 atmosphere is subjected to a potential difference of thousands of volts a cross a length of a few dm, by means of an induction coil or a transformer. A Geissler tube containing neon, even though it emits orange red light, can be grasped with hand.

In the spectral lamps with metallic vapours, the potential difference is a few tens of volts. In electro-luminescent condensers, a solid (zinc sulphide with traces of other metals) is incorporated into a plastic substance subject to an alternating field of the order of 10^7 V/m.

22.2   CHEMI-LUMINESCENCE

22.2.1   Principle

Various chemical reactions, principally oxidations are accompanied by an emission of light. The bio-luminescence phenomena (luciola, lampyris ...) are linked to chemi-luminescence.

22.2.2   Setting up of experiments

1) *Oxidation of phosphorus*

A piece of dry white phosphorus, placed in an atmosphere of oxygen, starts emitting light at about 30°C.

2) *Oxidation of luminol*

A small quantity of luminol[†] which is a yellow powder, is dissolved in 100 cm³ of a 1 % aqueous solution of potash. A few cm³ of hydrogen peroxide 100 volumes are added in the dark: a weak luminescence of light blue colour appears. A few drops of a concentrated solution of potassium ferricyanide are added: the luminescence becomes intense, without any increase in temperature.

3) *Oxidation of pyrogallol*

50 cm³ of each of the following aqueous solutions are mixed in a glass container of 300 cm³: pyrogallol at 10%; formaldehyde at 35%; potassium carbonate at 50%. The mixture is not luminous. 50 cm³ of hydrogen peroxide (100 volumes) are added in the dark; an orange red luminescence appears. Then the mixture gets slightly heated and froths.

22.3   CATHODE LUMINESCENCE

22.3.1   Principle

Electron bombardment of some materials may result in the emission of light. (The experiment of Franck and Hertz (21.4) is the simplest example of this category).

22.3.2   Setting up the experiment

1) Thoroughly evacuated glass bulbs fitted with two metallic electrodes may be used. Small solid fragments of different sorts (fluorite, calcite, ruby ...) are placed opposite to the cathode. On connecting the electrodes to an induction coil or a transformer, the luminescence of the solid fragments is observed under the impact of the cathode rays.

2) The screens of an oscilloscope or of a television receiver are common examples of cathode luminescence. The substance employed is zinc oxide (green), or zinc and beryllium silicates (yellow).

22.4   PHOTO-LUMINESCENCE

This is luminescence caused by the absorption of radiation. Fluorescence is different from phosphorescence in the following respects: fluorescence is said to occur when light is emitted within a short duration (10^{-8} to 10^{-1} s) after the exciting radiation has ceased to act and depends little on the temperature; in case of phosphorescence, the duration is generally higher (up to many minutes or even hours) and increases very much when the temperature is decreased. Only solids are phosphorescent. Photo-luminescence is distinguished from the diffusion of light (Rayleigh's effect and Raman's effect) by the fact that the exciting radiation should necessarily be absorbed.

Photo-luminescence obeys generally Stoke's law: the re-emitted radiation has a lower frequency than the exciting radiation.

22.5   OPTICAL RESONANCE AND FLUORESCENCE OF VAPORS

22.5.1   Principle

The molecules of a vapour may be excited by the absorption of an electro-magnetic radiation of suitable frequency. The spontaneous return of the molecule from the excited state to the fundamental state is accompanied by the emission of radiation.

If the emission transition $2 \rightarrow 1$ is inverse of the absorption transition $1 \rightarrow 2$ (fig. 22.1 a), the phenomenon is called optical resonance.

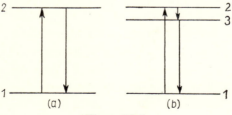

Figure 22.1

If the return to the fundamental state, 1, takes place through the inter-mediate levels $2 \rightarrow 3$, $3 \rightarrow 1$ (fig. 22.1 b), one or more of these transitions may be accompanied by emission of radiations having frequencies evidently less than the exciting frequency, this is the fluorescence.

22.5.2   Experiment on the optical resonance of sodium vapour (fig. 22.2)

The emission and absorption of the resonance line of sodium, $D_1 = 5896 \text{Å}$, takes place between the levels $3^2 P_{1/2}$ and $3^2 S_{1/2}$. (For the absorption, see § 20.2).

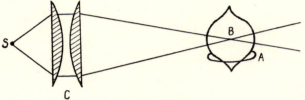

Figure 22.2

a) Apparatus

S: sodium vapour lamp emitting the yellow lines, or mercury vapour lamp with a filter to isolate yellow lines.

C: condenser converging the beam into the bulb *B* carried in a ring *A*.

B: bulb containing sodium vapour.

Fabrication of the bulb (fig. 22.3)

A clean and dry pyrex bulb, *B*, of 0.5 litre capacity caries an appendage *A* in which a small tube *T*, containing a piece of sodium the size of a pea, is

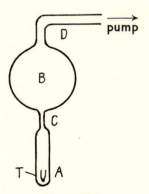

Figure 22.3

placed. A good vacuum (10^{-5} torr) is created with the help of a diffusion pump, and the glass walls (but not the sodium) are heated to eliminate the absorbed gas. The bulb is cooled, the process of evacuation is continued, and the sodium is heated slowly; it is distilled and is deposited on the walls of A. By successive heating, sodium is made to rise into the bulb B, on the walls of which it forms a brilliant deposit, The bulb is then sealed at C, and then at D.

b) Adjustments

The sodium has been collected at the lower portion of B, so that the incident beam can traverse the bulb. At normal temperature, the path of the beam remains invisible. B is then heated by playing the Bunsen flame on its sides. The vapour pressure of sodium rises, and the light path becomes illuminated brilliantly as a result of the emission of the resonance radiation in all directions. On removing the flame, the vapours condense and the emission decreases.

On interposing a shutter in the light path, it is shown that the emission ceases without appreciable delay (the mean life time in the excited state is of the order of 10^{-8} s).

The source is replaced by a mercury vapour lamp supplied with a filter which transmits the yellow lines only; the path of the light beam is not visible.

22.5.3   Experiment on the optical resonance of mercury vapour (fig. 22.4)

The absorption and emission of the resonance line $\lambda = 2537$ Å by an atom of mercury is produced by the transition between the levels 6^1S_0 and 6^3P_1, the fundamental level 6^1S_0 being normally occupied by two electrons.

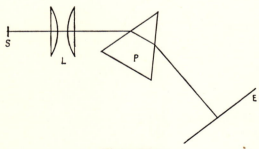

Figure 22.4

The resonance line being in the ultraviolet, its existence can not easily be demonstrated. The selective absorption of this line by mercury vapours is shown.

a) Apparatus

S: low pressure mercury vapour lamp with quartz envelope working on high voltage and emitting the resonance line ($\lambda = 2537$ Å) with great intensity. In some cases, a portion of the lamp is in the form of a linear capillary tube, which can serve as a quasi-linear source.

L: quartz lens, $f = 10$ cm.

P: quartz prism.

E: fluorescent screen (bristol paper).

b) Adjustments

The image of S is formed on E, and the orientation of P is so adjusted that the resonance line is at the minimum deviation position. The lines 3650 and 4047 Å (see the experiment 19.1) are easily observed. A cup containing a little of mercury heated to about 50°C, is placed between S and L; the resonance line disappears from the screen E but other lines of the spectrum remain.

22.5.4   Experiment showing the fluorescence of iodine vapour (fig. 22.5)

The fluorescent radiation of the iodine vapour has an orange colour. (Its spectrum is complex). It can be excited, for example, with the help of the green line of mercury. The phenomenon being quite weak, it must be observed individually, on a dark background, in a direction perpendicular to the direction of illumination.

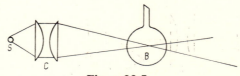

Figure 22.5

a) Apparatus

S: mercury vapour lamp, A_3 type, with a green filter.

C: condenser.

B: glass bulb containing a few crystals of iodine and sealed under vacuum.

22.6 FLUORESCENCE OF SOLUTIONS

22.6.1 Principle

Some substances in solution emit fluorescent radiation, which depending on their concentration, is visible. In water, fluorescein[†] and eosin[†]: yellow-green; rhodamine B[†]: orange; esculine[†] and quinine sulphate: blue.

In benzene, a solution of anthracene[†], or engine oil show a violet fluorescence. Visible fluorescence is excited by the "dark light" ultraviolet radiation (3651 Å) of mercury. The radiations which excite fluorescence are absorbed.

22.6.2 Experimental set-up

a) *Apparatus*

S: mercury vapour lamp emitting only in the ultraviolet (A_s).

C, C': glass containers with parallel faces.

D: diaphragm to limit the beam

b) *Adjustments*

1) A container, C, filled with distilled water is placed before the source (fig. 22.6). A few drops of one of the preceeding solutions are poured into C in the dark. The course of the liquid is revealed by the emitted fluorescent light.

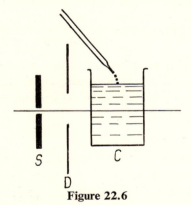

Figure 22.6

2) If S is replaced by a white light source, the fluorescent radiation emitted laterally by the fluorescin in the container C, is yellow-green. The transmitted light is yellow-orange, complementary to the blue which is absorbed.

3) The absorption of the radiation used for excitation can be demonstrated by placing another container, C', after C. Both contain quinine. The fluorescence appears in the first container but not in the second.

22.7 LUMINESCENCE OF SOLIDS

a) *Principle*

Some solids are luminescent when in the pure state: salts of the uranyl ion UO_2^{++} (yellow-green), calcium tungstate $CaWO_4$ (blue), salts of some rare earths, these of samarium (orange). Most solids owe their luminescence to a small quantity of an impurity called an activator: fluorite; ruby; silicates of zinc, calcium, and barium; zinc sulphide. Their photo-luminescence is employed for fluorescent illumination; their cathode-luminescence for oscillograph screens and for the television screens. One can find on the market red or orange papers impregnated with fluorescent material emitting yellow light under the action of violet or ultraviolet radiations. Fluorescent paints and colours are also available. Live teeth and nails also have a bluish fluorescence.

b) *Setting up an experiment to demonstrate the fluorescence of uranium glass*

The spectrum of a carbon arc is projected on to a screen (the intensity of illumination is of greater importance than the purity). A thick piece of uranium glass is moved across the spectrum: from red to green the uranium glass behaves as an ordinary glass; in blue and violet regions it emits green fluorescent light and gives a dark shadow on the screen.

22.8 DURATION OF LUMINESCENCE AND INTENSITY

The duration is very variable: zinc sulphides emit for many seconds; the salts of the uranyl ion and of rare earths emit for times of the order of 10^{-4}s. When the excitation is suppressed, the intensity of the luminescence decreases progressively.

A) *Measurements of these durations*

The durations can be estimated by placing the sample E (fig. 22.7), quite thin so as to be translucent, between two similar discs D and D'. Each disc has four radial windows, they are fixed on an axis A, in such a way that the windows of the two discs occupy alternate positions.

The whole system is rotated at a frequency N with the help of an electric motor of variable speed. The light beam used for excitation falls on E through the disc D. The luminescent light reaches the observer through D' after a time $1/8N$. By this method, one can easily recognize a difference of duration between the luminescence of calcite or of fluorite (10^{-1} s) and that of the uranyl ion salts.

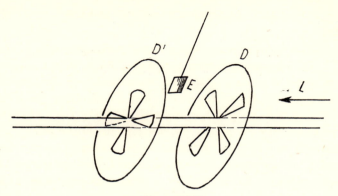

Figure 22.7

B) *Use of an electronic oscillograph**

The luminescent radiation illuminates a photo-electric cell and the current, produced as a function of time, is displayed on an oscilloscope.
Experimental set-up (fig. 22.8)

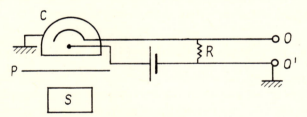

Figure 22.8

Apparatus

S: photographic flash lamp

P: luminescent powder (Massiot fluor, type 0132 Az) scattered on a glass plate.

C: photo-electric cell, evacuated.

V: 100-volt battery.

R: 1 MΩ resistance.

OO': terminals connected to the vertical plates of the oscillograph through a D.C. amplifier.

---

* This experiment has been conceived and set up by M. D. Fontaine, assistant in the Faculty of sciences, Paris.

Adjustments

The scanning of the screen takes about 0.5 s. *C* is illuminated with a flash in the absence of *P*. The decay curve has the shape of *A* (fig. 22.9). *P* is interposed and the experiment is repeated; the curve *B* is observed. All the wires should be shielded and earthed.

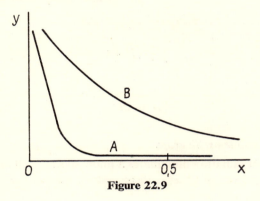

Figure 22.9

Formulae of
Organic Compounds Used

Anthracene: cyclic hydrocarbon, $C_{14}H_{10}$. Colourless crystals.

Bromophenol Blue: tetrabromophenol sulfone phthalein. Dye used to measure pH.

Methylene blue: tetramethylthionine hydrochloride, $C_{16}H_{18}N_3SCl$. Dye, dark blue powder.

Indigo carmine (or soluble indigo): $(C_8H_4ONSO_3Na)_2$. Sodium salt of disulphonated indigo. Dye, blue powder.

Ethyl cinnamate: $C_{11}H_{12}O_2$. Colourless liquid with a smell of cinnamon.

Cyanine (quinoline blue): $C_{25}H_{35}N_2I$. Dye in bronze-coloured crystals, used as a sensitizer in photography.

Eosin: tetrabromfluorescin. $C_{20}H_8O_5Br_4$. Dye, red powder.

Esculin: glucoside extracted from horse chestnut $C_{15}H_{16}O_9$. Colourless powder.

Fluorescein: sodium salt of resorcinolphthalein $C_{20}H_{15}O_0Na_2$. Dye, brownish powder.

Fuchsine: $C_{19}H_{18}N_3Cl$. Dye, red crystals which appear green in reflected light.

Luminol: hydrazide of amino o-phthalic acid $C_6H_7N_3O_2$, yellow powder.

Nitrobenzene: $C_6H_5NO_2$. Pale yellow liquid. For the study of electric birefringence, it must be distilled before use. The first fractions which contain water are rejected and what passes about 210°C is conserved.

Nitroso-dimethylaniline: $C_8H_{10}N_2O$. Green crystalline powder. M.P. 85°C (poisonous).

Pyrogallol (or pyrogallic acid): 1, 2, 3-trioxybenzene $C_6H_3(OH)_3$, colourless solid, generally oxidized into a brown one.

Rhodamine B: condensation product of phthalic anhydride with diethyl m-aminophenol. Dye, pink colour.

Methyl red: P-dimethylamino azobenzol-o-carbonic acid. Dye for pH measurement.

Preparation of Crystals and the Chemical Compounds Employed

All the crystals which can be prepared without difficulty are soluble in water. Those which are cited below, are prepared by slow evaporation of saturated solutions at constant temperature. The constancy of temperature is best obtained by working underground where the variations of temperature are seasonal, slow and of small amplitude (1 to 2°C). One may employ thermostats in which a higher temperature (25 to 40°C) can be obtained and a higher rate of evaporation; however, if there are interruptions in the mains electricity supply, the whole work has to be resumed.

In all cases, a solution supersaturated at the operating temperature is prepared. It is left as such for a number of days till thermal equilibrium is attained. The small crystals thus formed are separated by filtration, and the filtrate is seeded with one of them, chosen for its regular form. The solution S, so seeded and contained in a cylindrical glass vessel covered with a watch glass (fig. B 1), is left by the side of another analogous vessel containing a substance D (sulphuric acid or silica gel) which absorbs water vapours. The two vessels are placed under a bell jar whose ground and greased edge rests on a base plate B. The growth of the crystal seed is watched every day; it is replaced if it gets dissolved.

1) *Iodic acid* HIO_3

A solution containing 330 g of acid for 100 g of water gives large crystals by evaporation in a drying oven at 50°C.

2) *Potassium chlorate* $KClO_3$

The saturated solution at 20°C contains 7 g of salt in 100 g of water. The solubility is very much greater at higher temperatures. On allowing to cool slowly a solution prepared at the boiling point, or on evaporating a non-

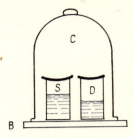

Figure B.1

saturated solution at ordinary temperature, a mass of small crystal plates is obtained; amongst these one can find brightly coloured crystals by examining the crystalline mass in reflected light after having decanted the mother-liquor. For spectral examination of these crystals, it is preferable to cement them, with Canada balsam, between two prismatic glass plates of small angle (fig. B.2).

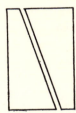

Figure B.2

3) *Sodium chlorate* $NaClO_3$

The saturated solution in 100 g of water contains 230 g of salt at 100°C and 100 g at 20°C. Clear cubes are obtained, the edges of which may reach many centimeters after 6 to 8 months.

4) *Copper and ammonium chlorides* $(NH_4)_2CuCl_4$, $2H_2O$

One mole of cupric chloride $CuCl_2$, $2H_2O$ and two of ammonium chloride NH_4Cl are dissolved in hot water and evaporated; blue quadratic crystals, are obtained.

5) *Sodium Nitrate* $NaNO_3$

The saturated solution at 15°C contains 85 g of the salt in 100 g of water. 100 g of salt is dissolved in 100 g of water by heating, then 20 g of pure tartaric acid is dissolved by heating and stirring. Clear and colourless rhombohedra are obtained which may have edges of many centimeters.

6) Sulphur

A solution of sulphur in carbon disulphide is slowly evaporated at room temperature by simply leaving the container opened.

7) Vanadium oxide V_2O_5

Colloidal solutions of V_2O_5 are prepared by grinding with mortar and pestle 0.5 g of ammonium vanadate and a little hydrochloric acid. The red precipitate is put on a filter paper or a sintered glass crucible and washed with distilled water till it starts passing through the filter. It is then placed in suspension in 200 cm^3 of distilled water and is stirred for a few hours. The dark red solution is filtered through paper.

8) Mercury-silver iodide Ag_2HgI_4

A little mercuric nitrate is dissolved in water to which just enough nitric acid is added to avoid the formation of the insoluble hydroxy nitrate. A solution of silver nitrate (1 atom Ag for 1 atom Hg) is added. A solution of potassium iodide, stirred continuously, is poured in the clear mixture. A bright yellow precipitate of the desired compound is formed which is separated by filtration, washed with cold water, with alcohol and with ether.

9) Dichroic liquid

A solution which has a marked circular dichroism for the yellow light of sodium can be obtained as follows: 0.75 g of potassium bichromate and 10 g of potassium tartrate are dissolved in 100 g of water and the mixture is heated to boiling temperature. The solution becomes green in colour; it is gently heated for a few hours to reduce its volume to about 20 cm^3. This solution, which can be kept indefinitely, is used for the experiments after cooling and suitable dilution (the dichroism diminishes slowly on dilution).

10) Colloidal solution of ferric hydroxide

A suitable ferric hydroxide sol can be obtained as follows: 10 cm^3 of a saturated aqueous solution of ferric chloride are diluted to 100 cm^3. Ammonia, which precipitates the hydroxide $Fe(OH)_3$, is added to the solution, stirred continuously, till the odour of ammonia is just perceptible in the mixture. This is centrifuged, the surface liquid is decanted and replaced by water in which the precipitate is again put in suspension.

The centrifugation is repeated thrice. Finally, the precipitate is put in suspension in 50 cm³ of water; 5 cm³ of the homogeneous suspension is taken off, diluted with 100 cm³ of water and three drops of a saturated solution of $FeCl_3$ added. After a number of days of stirring, the precipitate is peptised: the colloidal solution passes through the filter paper without a residue. It is heated in a water bath for at least three hours; its colour becomes more red and more intense. It is now ready for the experiment. By suitable dilution of the colloïdal solution, a compromise is to be found between the activity of the solution for the desired phenomenon, which increases with the number of particles encountered by light, and its absorption.

Cutting of Crystals

MICA

Mica plates for experiments on birefringence can be obtained without the help of specialists. Mica muscovite originating from Brasil or India is brownish in colour when its thickness is of the order of a centimetre. It can be cleaved easily by starting a cleavage with a razor blade or a needle and continuing it either by introducing in the slit a bristol paper (visiting card) with a chamfered edge, or by separating the two parts with hands, in distilled water contained in a crystallising vessel. This mica is a negative, biaxial crystal. The neutral lines of the cleaved plate have refractive indices $n_m = 1.593$ and $n_g = 1.597$ (for the D line of sodium).

The half-wave plates for the visible have, thus, a thickness of the order of 70 μ. To choose birefringent plates of known retardation we proceed as follow:

1) Regions of uniform thickness are looked for in cleaved plates by placing them in front of a large monochromatic source and examining them in reflected light (§ 3.2.4). Fringes of equal inclination are observed; their spacing and position changes abruptly when the thickness of the plate changes. The regions of uniform thicknesses are encircled by a line in ink.

2) The retardation δ is determined approximately by means of the phenomena of chromatic polarisation (n° 14.1). Nörrenberg's apparatus (fig. C.1) is very useful for this purpose. Natural white light emitted by a big incandescent lamp S (with frosted bulb) falls under Brewster's incidence on a glass plate G (un-tempered) which sends the polarized light vertically towards the lower end of the apparatus. The light beam is reflected normally by the mirror M, passes through G and a small piece of polarizing sheet placed before the eye. Light can be extinguished by rotating A. The plate under

study is placed at (1) on a diaphragm and its hue is observed; A is then rotated through 90° to observe the complementary hue. This operation is repeated with the plate at (2): the retardation is approximately 2δ. There are thus 4 hues which permit the use, without ambiguity, of Newton's scale of colours.

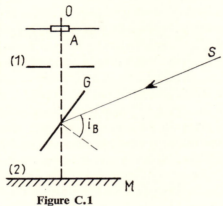

Figure C.1

The sensitive hue (tint of passage) of the first order will be produced by a plate placed at (1): if it is a full wave plate (for the mean yellow) and is placed between crossed polarizers, or if it is a half wave plate and is placed between parallel polarizers. With the plate at position (2), the sensitive hue is obtained when it is a quarter-wave plate, and is placed between parallel polarizers, or when it is a half-wave plate and the polarizers are crossed. Instead of the apparatus illustrated in fig. C.1, it is sufficient to arrange the following elements in a vertical frame (fig. C.2): an incandescent lamp L, a ground glass V, two polarizing sheets P and A of at least 5 cm on each side and of which at least one can be rotated in its plane. The plate to be examined is placed at P. In this arrangement the advantage of double passage is not obtained as in the Nörrenberg apparatus.

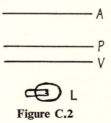

Figure C.2

3) The visible radiation for which a plate is a half-wave can be determined by examining its channelled spectrum between parallel polarizers using the set-up of figure 14.10 in which the direct vision prism should be replaced by a calibrated spectroscope. The wavelength corresponding to the single dark band is measured.

4) The visible radiation for which a plate is a quarter-wave can be determined by the preceeding method, the light traversing the plate twice under almost normal incidence (using, for example, the arrangement of figure C.2). The single polarizing sheet P polarizes the incident beam and analyses the reflected beam.

5) To determine the directions of neutral lines of a plate L, it is placed in the arrangement of figure C.2 between crossed polarizers. The plate is rotated in its own plane till extinction is obtained. Another mica plate L' of arbitrary thickness and with a straight edge is superimposed on the former plate (fig. C.3). L' is rotated to obtain extinction and a line B_1 is drawn

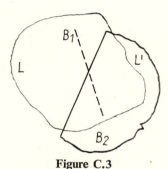

Figure C.3

on L along the straight edge of L'. The plate L' is turned so that its two faces interchange their positions; it is now rotated in its own plane to obtain extinction and a line B_2 is drawn on L along the straight edge. The directions of the bisectors of the internal and external angles between B_1 and B_2, determine the neutral lines of L. If a plate L' is prepared in this way, it can be cut along a straight edge parallel to one of the neutral lines; the line B_1 in the preceeding operation determines a neutral line of the plate L.

6) The neutral line corresponding to the refractive index n_m (fast axis) is determined by using the fact that this direction is normal to the plane of

18 a   Francon

the optic axes $n_p n_g$ (fig. C.4). The plate is placed in the apparatus illustrated in figure C.2 and is inclined about a neutral line; if this neutral line has the index n_m, the retardation decreases (this decrease is shown by a variation in the hue) and becomes zero when the light propagates along any one of the axes A or A'. If the plate is inclined about the neutral line corresponding to n_g, the retardation increases constantly.

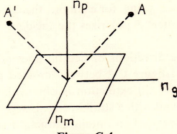

Figure C.4

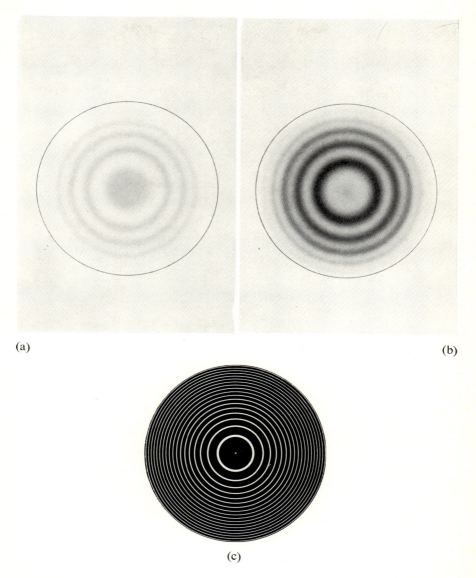

(a)

(b)

(c)

Plate I (a) Newton's rings by transmission. (b) Newton's rings by reflection. (c) Fabrey-Pérot fringes

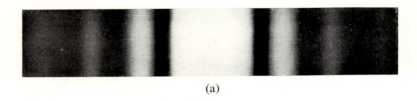

(a)

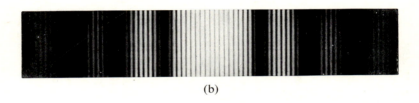

(b)

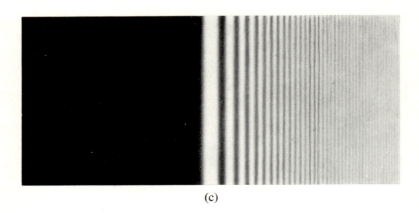

(c)

Plate II (a) Diffraction pattern of a narrow slit. (b) Diffraction pattern of two identical and parallel narrow slits. The diffraction pattern is modulated by the interference phenomenon produced by the two slits. (c) Diffraction fringes near the edge of an opaque screen

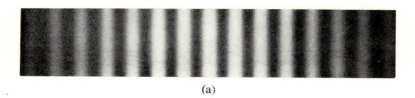

(a)

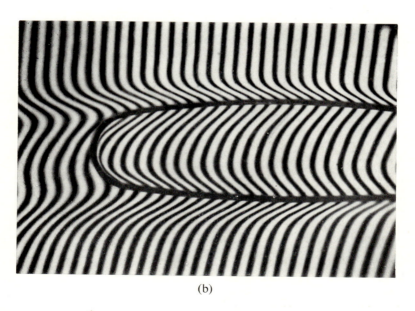

(b)

Plate III (a) Young's fringes. (b) Michelson interferometer. Fringes of an air-wedge deformed in the vicinity of a wire heated by an electric current.

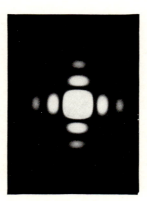

Plate IV (a) Fringes of equal thickness of a variable thickness thin plate.
(b) Fraunhofer diffraction by a square aperture

Index

18 a*